CONSUL • ZEPHYR • ZODIAC

Dedication

This book is dedicated to the memories of Raymond Mays and T. C. 'Cuth' Harrison, both of whom played an important part in the competition successes of the Ford Zephyr.

Reproduced courtesy of Autocar.

CONSUL • ZEPHYR • ZODIAC

The Big Fifties Fords

MICHAEL ALLEN

MRP

MOTOR RACING PUBLICATIONS LTD
28 Devonshire Road, Chiswick, London W4 2HD, England

ISBN 0 900549 80 7
First published 1983

Photosetting by Zee Creative Ltd., London SW16
Printed in Great Britain by Netherwood, Dalton & Co. Ltd., Bradley Mills, Huddersfield, West Yorkshire

Contents

Acknowledgements		6
Chapter 1	Dagenham after the War Anglia, Prefect and Pilot	7
Chapter 2	The Five Star Cars Mk 1 Consul, Zephyr and Zodiac	10
Chapter 3	The Three Graces Mk 2 Consul, Zephyr and Zodiac	57
Chapter 4	Racing and rallying The Big Fifties Fords in competition	93
Chapter 5	Obsolescence and renaissance From bargain basement to collectible classic	130
Chapter 6	Owning a Big Fifties Ford Performance, purchase and maintenance	137
Chapter 7	A very special Zephyr Case history of Peter Scott's car	147
Chapter 8	The Ford-powered sports cars Allard, Lea-Francis, Reliant and others	152
Appendix A	Technical specifications	156
Appendix B	Colour schemes	156
Appendix C	Production figures	157
Appendix D	Major competition awards	158
Appendix E	Lubrication schedule	160
Appendix F	Clubs and spares specialists	160

Acknowledgements

It would not have been possible to write this book without the help and encouragement of many people, and I must therefore express my gratitude to the following: David Burgess-Wise, of Ford Motor Company Public Affairs, and Ford Press Officer Derek Sansom, for their help and kindness when I visited Brentwood; Ford Personnel Supervisor D. J. Pasterfield, and former Ford employees Les Geary and George Halford; Mk 1 Owners' Club founder Dave Barry, and Mk 1 Club Secretary Dick Barry, for their help, particularly with the 'Australian connection'; Mk 2 Owners' Club Secretary and founder member Dave Debenham, and Mk 1 and Mk 2 Club members Howard Foottit, Eric Pearson, Stuart Clarke, Alan Poole, Ray Greenaway, David McNiece, Dave Cropper and Keith Roe, who between them came up with a variety of assistance, including the loan of literature and the tracking down of others who could perhaps help answer vital questions; Peter Scott, the most remarkable Mk 1 Zephyr enthusiast I have ever met, who supplied a wealth of information about his own very special Zephyr; obsolete Ford spares specialist Ken Tingey, and my two mechanic friends Allan Marshall and Michael Mills, both of whom in addition to answering questions have also looked after a succession of Zephyrs of my own over the years; A. H. Leake, of Laycock Engineering, J. N. Green, of Servais Silencers, and the Leeds Divisional Sales Office of the Michelin Tyre Company; *Autocar* Editor Ray Hutton, and *Motor* Editor Tony Curtis, for their permission to use extracts from road tests and articles which have appeared in their respective magazines; John Bolster, of *Autosport*, for his permission to quote from his road test reports, and Simon Taylor, Managing Director of Haymarket Publications, the proprietors of *Autosport;* the Australian motoring magazines *Wheels* and *Cars Today;* Peter Deffee and Eric Walker, of the Classic Saloon Car Club; Miss Patricia Hulse, of the Allard Owners' Club; Martin Ashman and Keith Tricker, of the Lea-Francis Owners' Club; and A. B. Price, of Lea-Francis Cars.

Racing and rallying played an important part in the story of these cars, and in this respect I have been fortunate enough to have made contact with many of the personalities involved. Those who were within striking distance invited me to meet them, and my sincere thanks for their willing help and such kind hospitality go to Anne Hall, Edward and John Harrison, Denis Scott, Ken Armstrong and Jack Reiss. For their informative letters, and their patience at the other end of the telephone, I am also most grateful to Jeff Uren, Gerry Burgess, Eric Jackson, former Ford Competitions Manager Edgy Fabris and Competitions Department mechanic Norman Masters. I would also like to thank Mike Wilds, of the BRSCC, for additional information.

Many of the illustrations have been supplied by the Ford Motor Company, and for these I would like to thank Steve Clark and Sheila Knapman, of Ford Photographic Services. Others have come from several sources, including *Motor*, and I thank the Editor for his permission to use these, and his secretary, Sally Crompton, for unearthing the old negatives. Thanks are also due to Barry Hughes, of the Reliant Motor Company, and the aforementioned A. B. Price of Lea-Francis Cars. The remainder of the illustrations are from private collections, and so here once again my thanks go to Anne Hall, Patricia Hulse, Edward and John Harrison, Jeff Uren, Denis Scott, Ken Armstrong, Edgy Fabris, Norman Masters, Eric Walker, Keith Tricker, Jack Reiss, Peter Scott and Dick Barry.

Finally, I must thank my mother, for all her encouragement and literary tuition, and, of course, MRP publisher John Blunsden, for his faith in my enthusiasm for Consuls, Zephyrs and Zodiacs.

June 1983 Michael G. D. Allen

CHAPTER 1

Dagenham after the War

Anglia, Prefect and Pilot

As the Second World War receded into the past, the British motor industry, which had worked so hard throughout the period of hostilities turning out an amazing variety of armaments and military vehicles, was able to turn its attention once again to the production of passenger cars. Almost inevitably, all its early offerings were nothing more than thinly disguised versions of the manufacturers' respective prewar models, hastily re-introduced with the minimum of revision to either the styling or the mechanical specification, and intended only to fill in the time that it would take for the design and engineering teams to come up with something entirely new.

So by the end of the 1940s a steady stream of new models was emerging from the factories of most of the major manufacturers, often featuring overhead-valve engines, independent front suspension and, in some cases, monocoque construction that did away with a separate chassis-frame, and all displaying styling features which immediately rendered visually obsolete their prewar-inspired predecessors. Meanwhile, to the surprise of many people, the production lines at Dagenham continued to turn out what were barely facelifted prewar Fords, and it appeared that the Ford Motor Company, who had so often been the market leaders in the past, were now being left well and truly behind.

At the economy end of the range, the familiar 8hp 933cc two-door Anglia and the 10hp 1,172cc four-door Prefect were selling well, their 1930s ancestry allowing the Ford Motor Company to list them in 1948 at £293 and £352, respectively, and so fully uphold the Ford tradition of value for money.

The first serious opposition to the small Fords came in October 1948 with the introduction of the Morris Minor, a model that set entirely new standards of road behaviour for this class of car, and was also extremely competitive at a price of £358. It was, however, let down somewhat at this stage by the rather surprising use of a prewar side-valve engine that itself had Ford connections. When, in 1932, Dagenham had introduced the all-new 8hp Y type Ford and, at a stroke, rendered obsolete such opposition as the contemporary Morris Minor and Austin 7, Morris Motors had purchased an example of the new Ford, taken it to pieces, carefully measuring the components as they went along, and then proceeded to produce a remarkably similar version of the engine for their own new car,

The V8 Pilot offered excellent performance allied to a very high standard of finish and equipment, but the running gear and the appearance of the car were totally out of date by 1950.

which had been announced in 1934 as the Morris 8. It was a developed version of the engine from this car that was used to power the otherwise completely new 1948 Morris Minor.

Other than the Minor there was little opposition to the small Fords at this time, as the other rival manufacturers had been concentrating their efforts on new models in the medium-size category, and so it was considered at Dagenham that with their firmly established following, coupled with the low prices at which they were being offered, the Anglia and Prefect could soldier on into the 1950s.

Above the Prefect, however, there was a considerable gap in the Ford range, and those motorists who wished for something in the $1\frac{1}{4}$- to $1\frac{1}{2}$-litre category, with a modern appearance and an up-to-date mechanical specification, were placing their orders for the recently introduced Austin Devon and Dorset, Morris Oxford, Hillman Minx and Vauxhall Wyvern, models that were well suited to satisfy the demands of what was to prove a rapidly expanding sector of the car buying public, which was not being catered for at all by Ford at that time.

Ford of Dagenham's flagship for the late 1940s was the big, V8-engined Pilot. A fast, roomy and very well appointed car, it could be regarded as the logical conclusion of a sturdy theme whose origins could be traced back to 1932, when Henry Ford had triumphantly announced to the world the first Ford V8. With its cylinder block cast in one piece, this engine had been regarded by many contemporary observers as something of an engineering miracle.

Priced at £764, the Pilot, like its smaller stablemates, represented typical Ford value. The 3.6-litre side-valve V8 engine propelled the 30cwt car with remarkable ease, giving a maximum speed of more than 80mph under favourable conditions, whilst at the other end of the scale allowing a driver to retain top gear right down to below 10mph and still be able to pull away from this speed smoothly and quite rapidly. Contemporary road test reports published in the popular motoring press revealed that the Pilot had the ability to reach 70mph from a standing start in around 30 seconds if full use was made of the three-speed gearbox. By the standards of the day all this was, of course, very definitely 'high performance', but the price was a

The 1949 American Ford on which the styling of Dagenham's Consul and Zephyr Six was based.

fuel consumption that could be as high as 14mpg in heavy traffic conditions, and even if a feather-footed technique was adopted whilst driving on the open road, the result was unlikely to be anything better than 19-20mpg. This was less than satisfactory in the austerity of those early postwar years.

The suspension, with its transverse leaf springs front and rear, was very much 'traditional' Ford. Aided by the excellent seats, it would give the occupants of the Pilot a very comfortable ride, but in terms of handling, the big Ford could not hope to match the agility of some of the more recent offerings from rival manufacturers with their newly-adopted independent front suspension systems. The design showed its age, too, in the use of a braking system with hydraulic operation for the front brakes only, whilst the rear shoes were activated mechanically, resulting in high pedal pressures being called for, particularly if an emergency stop was required.

A close examination of the fittings and furnishings of the Pilot would reveal a high degree of quality and good workmanship, and in this respect particularly, the Pilot was a product of which Dagenham could be justifiably proud. But with large separately-mounted headlamps and wide running boards accentuating its typically 1930s styling, as well as its dated mechanical specification, it was obvious that it could not hope to compete for long in the postwar market place.

Clearly, if Dagenham were to take a lead in the 1950s, they urgently needed a new car that would not only replace the big V8 and perpetuate its high performance with a new-found economy, but would also fill that important gap in the middle of the range. In addition to all this it would have to be sufficiently advanced in both style and engineering to ensure a lengthy production run and, of course, bearing the name of Ford, it would have to sell at a bargain price.

CHAPTER 2

The Five Star Cars

Mk 1 Consul, Zephyr and Zodiac

Ever since the assembly lines at Dagenham had first come into operation in October 1931, the cars which emerged from the Thameside factory had shown a strong American influence. The diminutive 8hp Y type of 1932 had been wholly designed in Dearborn, and the big V8 models that had culminated in the postwar Pilot could be regarded as little other than right-hand-drive versions of a typical late 1930s full-size American Ford.

In the United States, passenger car production was sanctioned once again by the Government soon after VE day and, characteristically, it was the Ford assembly lines at Dearborn that were the first to roll. On July 3, 1945 the first postwar Ford, a Tudor (two-door) Super De Luxe left the line and was personally delivered by Henry Ford II, grandson of the famous founder, to US President Harry S. Truman.

Owing to the untimely death of his father, Edsel Ford, young Henry had been 'thrown in at the deep end' and had had little time to gain the experience necessary to hold down the Presidency of an industrial giant such as the Ford Motor Company. Wisely, he looked around for outside help, and in 1946 he recruited Ernest R. Breech, former President of the Bendix Aviation Company, to be his Executive Vice-President at Ford.

As was the case at Dagenham, the new cars being assembled at Dearborn were virtually identical with the last of the prewar models, and Breech decided that an all-new model would be needed quickly if Ford were to re-establish their position as leaders in the low-price field. With the full agreement of Henry Ford II, he set the engineers to work on a completely new chassis which would do away with the traditional Ford running gear that had its origins in the famed model T. In order to ensure that the new car would also have a completely new image, an outside styling consultant, George Walker, was brought in to produce the mock-ups for the body. The new car that was soon taking shape as a result of these decisions was appreciably bigger and heavier than Dearborn's previous products, and although the chassis was to feature entirely new running gear with wishbone-type independent front suspension and longitudinally mounted leaf springs at the rear, the car would still be powered by the venerable side-valve V8 engine and, surprisingly, would continue to utilize a 6-volt electrical system.

In 1948, a team of engineers from Dagenham visited the parent company in Dearborn to discuss plans and proposals for a new British

Ford. It was obvious from the outset that the new car nearing completion at Dearborn would be far too big and heavy, and therefore also too thirsty, to stand a chance of significant sales in postwar Britain or Europe, or in the many other world markets in which Dagenham competed so successfully, and so a right-hand-drive version of this particular car for Britain was never under serious consideration. It was important, however, that the new image that was being created by stylist George Walker should be a feature of the Dagenham product if at all possible, and fortunately the final shape which had evolved at Dearborn lent itself well to the scaled-down dimensions that would be acceptable in Europe.

Executive engineer George Halford was a member of the Dagenham team, and he recalled later that the 'joint consultations' which took place with their American hosts went very smoothly, with no serious areas of disagreement, and as a result of these talks, the team returned to England with plans for a car that would be technically far in advance of the new American Fords. A completely new four-door bodyshell, incorporating a large glass area and of monocoque construction, had been decided upon. Development of this new body was to be in conjunction with long-time suppliers of Ford bodywork, Briggs Motor Bodies, whose subsidiary at Dagenham had been established in the 1930s at the suggestion of Henry Ford in order to enable them to supply his English operation. Briggs had already some experience of monocoque construction, as the parent company had supplied prototype unitary bodies of their own design to the Lincoln Division of the Ford Motor Company for the 1935 Lincoln Zephyr.

Whilst featuring quite substantial chassis-type members welded into its

Consul and Zephyr Six, the 'Five Star Cars'. According to the brochure, the five star features were: OHV 'oversquare' engines; 'centre slung' within-the-wheelbase seating; independent front suspension; all-steel, integral body/chassis construction; hydraulic brakes all round.

lower regions, the new bodyshell would also use its large side frames, the front scuttle/toeboard bulkhead and, through the windscreen pillars, the roof pressing in combination with each other to provide a structure of immense strength and rigidity without excess weight. The rear extension of the bodyshell formed an unusually large full-width luggage compartment, under the floor of which was situated a 9-gallon fuel tank, protected at the sides by the chassis rails and at the rear by the bumper bar, the mountings of which were themselves extensions of the chassis members. The upper halves of the rear wings were part of the integral construction, whilst the lower halves were unstressed bolt-on panels. Forward of the windscreen the bodyshell consisted of sturdy inner wing panels, and across the front end joining these together was another welded-in bulkhead featuring a large central aperture into which the radiator would be bolted. Additional bracing of the front end structure was by struts welded into the inner wings at the point where the suspension upper mounting was situated and running diagonally across the engine compartment to be welded into the scuttle beneath the windscreen. The whole front end structure differed in length according to which of the four- or six-cylinder engine options that had been decided upon for the new cars was to be installed. The front wings, which contained the lamp units, were bolted on, as was a separate front panel. This latter section varied to accommodate the different grilles that would identify the four- or six-cylinder model.

One of the factors that led the design team to choose monocoque construction for the new cars was that one of the American Ford engineers, Earle MacPherson, had recently schemed out a brilliantly simple independent front suspension system for use with an integrally-constructed body. This front suspension, which eventually became known universally as the MacPherson strut and would also, after the Ford patents on its design had exceeded their time limits, be widely adopted throughout the motor industry, was built up around what was in effect a very long

Consul and Zephyr Six MacPherson-strut front suspension, looking forward. The tubular crossmember in the centre bolts to the chassis rail along the lower edge of the inner wings. When installed, the steering box is situated above the track rods (lower right in this illustration) and is thus behind the axle line, a valuable safety feature.

Rear end arrangement. Particularly noteworthy is the five-stud fixing for the road wheels. Four studs are considered adequate for this weight of car today, and Renault use only three studs on their similarly-sized models.

kingpin, the upper part of which was in the form of a telescopic shock absorber. This section was surrounded by a coil spring and was mounted at the top to the inner front wing, and so, as the inner wing was an integral part of the whole shell, the stress was widely spread, with the front scuttle/bulkhead, the floor and, via the windscreen pillars, the roof, all combining to take the load. By utilizing the strongest part of the car in this way, the MacPherson strut offered substantial savings in both cost and weight over the conventional subframe-mounted double-wishbone installation. At the bottom, a track control arm linked the suspension strut to the front bodywork crossmember, and was thus effectively one half of a lower wishbone, the other half of which was the swept-back end of a transverse anti-roll bar.

This was, of course, the first application of the MacPherson strut, and trouble was experienced with too much friction in the system in the early prototype cars, although, according to engineer George Halford, this was successfully overcome before the production stage was reached by very careful attention to detail. At the rear the car was supported by longitudinally-mounted leaf springs, on which was carried the rear axle, controlled by lever-arm shock absorbers. The steering gear chosen utilized a Burman worm-and-peg type steering box, which, in company with the new front suspension, gave the cars a far more accurate degree of control than on any previous Dagenham product.

The braking system was by Girling, and consisted of 9in diameter drums all round, those at the front being of the two-leading-shoe arrangement. The total lining area was 121sq in, and the system was activated hydraulically from a master cylinder mounted high up on the scuttle above the toeboard, operated by a pendant brake pedal. The mechanically-operated handbrake activated the rear shoes only, controlled by a pull-out lever under the dashboard.

The 13in diameter road wheels were shod with 5.90 x 13 tyres on the four-cylinder car, running at a recommended pressure of 28psi, and on the six-cylinder model the very amply sized 6.40 x 13 on $\frac{1}{2}$in wider rims, running at 24psi pressure. The track was 4ft 2in at the front and 4ft 1in at the rear, whilst the wheelbase was 8ft 4in and 8ft 8in, respectively, for the four- and six-cylinder models.

It was in order to achieve both the performance and the economy levels that would appeal to the widest possible sector of the buying public that two engine capacities had been decided upon for the new range. An in-line or 'straight' six-cylinder engine was chosen for the high-performance model, as six cylinders in this configuration, driving a three-plane crankshaft, offer outstanding mechanical smoothness accompanied by a very worthwhile saving in cost and weight in comparison with a V8 engine, and can in addition be very conveniently reduced in length to four cylinders, and so provide another engine of two-thirds the capacity with the maximum of interchangeability of internal moving parts between the two units. The abolition of the 'horsepower' tax in Great Britain in 1947 had been very timely as far as the Ford engineers were concerned. This tax, which was calculated on the area of the piston crown, had forced engine designers to use narrow-bore engines with a long piston stroke in order to be in the lowest practicable taxation class with any given cubic capacity, and had thus hampered engine development for many years. This narrow-bore, long-stroke arrangement was undesirable as it resulted in very high piston speeds at relatively low rpm, and also required a long-throw crankshaft, which can be lacking in rigidity. Freed from these constraints, the Dagenham engineers were able to adopt wider bore, shorter stroke dimensions, and produce what was to become known as an 'oversquare' engine, 'square' being the term used when the bore and stroke dimensions are identical.

The weight-saving process with the six-cylinder version was apparently taken too far at first as, according to George Halford, the original prototype cylinder blocks proved to be too thin, and a new block with considerably more 'meat' was introduced early in the development programme. To ensure even cooling, the water jacketing circled around each individual cylinder, and this, coupled with the large bore, resulted in a somewhat long engine in six-cylinder form. The cylinder dimensions chosen, 79.37mm bore x 76.2mm stroke, (3.125in x 3.00in) gave a capacity of 1,508cc (92cu in) and 2,262cc (138cu in), respectively, for the four- and six-cylinder units. This short piston stroke required only a very

Cutaway view of the left-hand side of a Zephyr Six engine. Water jacketing around each cylinder in a large-bore 'six' results in a long engine, but ensures even cooling and freedom from distortion problems. The cutaway of the right-hand side of the engine shows camshaft and pushrod details. The air cleaner is the oil bath type fitted for certain export markets where dusty conditions prevailed. A sump guard and a four-bladed fan were also available in certain territories.

short-throw crankshaft in comparison with earlier designs, and with a crankpin journal diameter of 1.93in allied to a main bearing journal diameter of 2.25in, the cast-alloy steel shaft was an extremely rigid affair even in its comparatively long six-cylinder form. Fully counterbalanced, the crankshaft ran in either three or four main bearings depending upon the four- or six-cylinder application. These bearings, along with those of the connecting rod big-ends, were steel-backed with a white-metal lining, and were of the replaceable shell type.

Overhead valve gear, Zephyr Six. The pipe to the centre of the rocker shaft is the oil feed. The three studs are for rocker cover fixing.

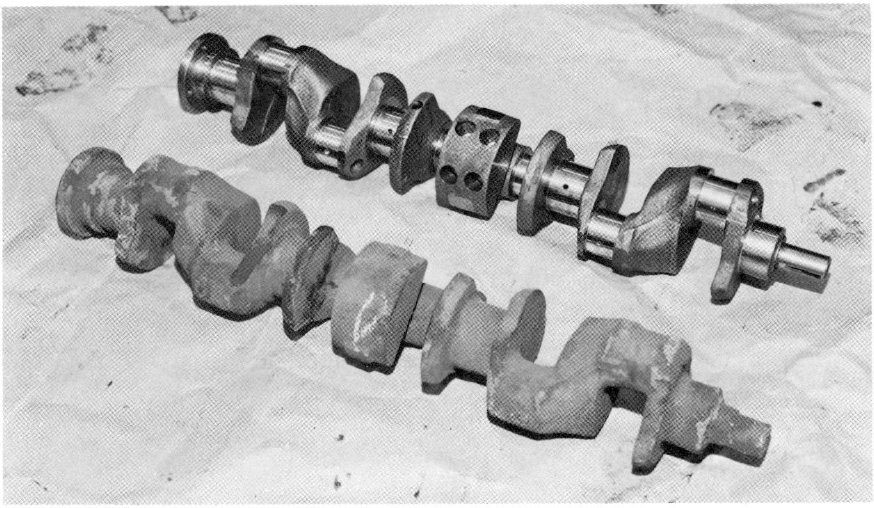

Zephyr Six crankshaft in both original cast state and fully machined. The extremely short throw, allied to wide bearing journal diameters, can be appreciated in this view.

Zephyr Six engine, showing the simple exhaust manifold.

Consul engine with standard, home-market air filter. The tube from the side cover is the crank-case breather. Note the gearbox dipstick which is reached from inside the car by pulling back the floor covering and removing a small plate in the transmission tunnel.

Aluminium alloy split-skirt pistons of the autothermic type were used, fitted with two compression and one oil control ring above the gudgeon pin, and the connecting rods were steel forgings of I-section. Lubrication was by a gear-type pump situated in the sump and driven from the camshaft; a full-flow oil filter with a large replaceable element was incorporated in the system. Overhead valves were another innovation for Ford, and were operated by what could be described then as the 'conventional' arrangement of pushrods and rockers. The valves opened into wedge-shaped combustion chambers in the cast-iron cylinder head, which featured siamesed inlet ports and gave a necessarily low compression ratio of 6.8:1 in order to cope with the very low octane petrol of the period.

A camshaft-driven AC mechanical fuel pump supplied petrol to a single Zenith carburettor in both cases. The cast aluminium inlet manifold had very carefully thought out internal passages on the six-cylinder application, designed to swirl the mixture and so ensure that each cylinder would receive an equal amount of fuel despite the engine's length. In contrast, the exhaust manifold seemed a particularly primitive affair, consisting of a pipe simply clamped to the side of the cylinder head and designed to take the gases to the front of the engine before curving back into the downpipe and thence into the silencer. The cooling system, pressurized to 4psi by the

Classic layout. An excellent illustration of the straightforward simplicity of the 'Five Star Cars'. These models exhibited a combination of features which would eventually be copied by many other major car manufacturers.

radiator cap, was pump-assisted and of $16\frac{1}{2}$ pints or 22 pints capacity according to the engine size.

The four-cylinder engine developed 47bhp at 4,400rpm and 74lb/ft torque at 2,400rpm. The corresponding figures for the six-cylinder unit were 68bhp at 4,000rpm and 108lb/ft at 2,000rpm. With the low piston speed of 500ft/min/1,000rpm these engines had a built-in capacity for sustained hard work coupled with long life.

The 8in diameter clutch was hydraulically operated for the first time on a Dagenham car, its master cylinder being situated alongside that of the braking system, operated by a similar pendant pedal and sharing the same fluid reservoir. This reservoir had an internal division, so that a leak in either circuit would not affect the operation of the other. The bhp and torque characteristics of the new engines were well suited to a three-speed gearbox, and so by retaining this traditional Ford layout, complete with an unsynchronized first gear, worthwhile savings could be made in the overall development costs of the new models. The gears were selected by a steering column-mounted lever which was chosen in order to allow unrestricted access to the front bench seat. An open propeller shaft replaced Ford's previous torque tube arrangement, and transmitted the power to a new hypoid-bevel final-drive assembly. The final-drive ratios chosen, of 4.625:1 and 4.375:1, respectively, for the four- and six-cylinder models, gave relatively low overall gearing that would endow the cars with exceptionally good top gear acceleration and, because of the short-stroke engine's willingness to rev freely, would still allow useful maximum speeds of around 73-74mph and 80-81mph, respectively, at approximately 5,000rpm, but at only 2,500ft/min piston speed. Typical rival designs of similar engine capacities, but of the then-normal long-stroke arrangement would reach these piston speeds at around 60-65mph or, in a few cases, even less than this. So, by comparison, the new Ford engines would be operating under far less stress, even when being driven at maximum speed.

A 12-volt electrical system was another departure from previous Ford practice, although in spite of this, and the provision of a very ample battery, the windscreen wipers were still vacuum-operated from the inlet manifold in the time-honoured Ford tradition. These wipers did, however, have the advantage of a vacuum booster pump which was incorporated in the base of the fuel pump, and if the wiper control was set towards the slow end of the motor's variable speed range, the booster pump ensured that the previously annoying tendency of the wipers to stop under wide throttle openings was largely overcome.

During the early development the new cars were known only by their factory designations, EOTA and EOTTA for the four- and six-cylinder versions, respectively. Before the production stage was reached, the name Consul was applied to the four-cylinder car and, as the Lincoln Division of the Ford Motor Company in America had decided to phase out their low-priced Zephyr model in order to concentrate on the luxury Continental, Dagenham adopted the Zephyr name for the six-cylinder model. The Lincoln Zephyr had always been powered by a V12 engine, and in order to avoid any possible confusion with that unit, the full title 'Zephyr Six' was to appear on the Dagenham car.

As the rear seat was well within the wheelbase, a good, level ride was assured, and with a width of 50in and 53in for the front and rear bench seats, a six-seater claim was entirely justified. Access to the interior was excellent through the wide-opening front-hinged doors. Rather in keeping with what was Dagenham's established image, the interior appointments were not particularly lavish in the new cars, the designers having gone for a simple, functional durability at this stage. Upholstery was in PVC, although leather would be available as an optional extra, and both front and rear seats were devoid of a central armrest. Both cars featured swivelling quarter-lights in the front door windows. The Zephyr Six had padded armrest-cum-doorpulls on all four doors, but these were omitted from the Consul, the floor of which was covered by a moulded rubber mat, whereas the Zephyr Six featured fully fitted carpets with bound edges. Both cars had cloth roof linings, and the twin sun visors were covered in the same material. The interior light, mounted centrally in the roof, was not automatically illuminated by opening the doors. Two powerful horns were sounded by means of a chrome-plated full-circle horn ring attached to the T-spoked steering wheel on the Zephyr Six; the Consul's single horn was operated by a button in the wheel centre. Also mounted centrally in the steering wheel boss was the self-cancelling switch for the semaphore-type direction indicators. Facing the occupants was a simply-styled flat dashboard incorporating, in front of the driver, a speedometer with total mileage recorder, ammeter and fuel contents gauge. Warning lights were provided for ignition and low oil pressure. In front of the passenger was a small lidded compartment which matched in shape the instrument surround, and mounted centrally along the lower edge of the dash panel were the choke knob and the switches for the starter, ignition and lights, panel light and wipers. These switches, along with the gear lever knob, were

1951 Consul with optional radio. When fitted by the factory or a Ford dealership, the radio aerial was always placed in the position shown, which allowed it to be extended by the driver without leaving the driving seat.

of an attractive 'see through' design. A fresh-air heater and demister unit, complete with a booster fan, was an optional extra, and when fitted its controls were situated beneath the dashboard panel in a central position.

The spare wheel was housed in an upright position in the offside of the luggage boot and could be extracted without having first to remove any luggage that was being carried. Of 15cu ft capacity, the boot was very conveniently shaped, and the fully counterbalanced lid was a nice finishing touch. At the front, the large bonnet was held in the open position by a ratchet-type stay, and accessibility for routine maintenance was excellent; the dipstick, oil filler, hydraulic reservoir, distributor and battery were all ready to hand. In contrast, the fuel filter was rather difficult to reach as it was incorporated in the fuel pump body, which was mounted low down on the side of the engine and was partially obscured by the optional heater unit, which few Consuls or Zephyrs, in fact, seemed to be without.

Exterior chromium ornamentation was confined to the grille, bumpers, hub caps, door handles, lamp surrounds, wiper arms and name badges on the Consul. In addition to these items the Zephyr Six featured chrome-plated bumper overriders, window surrounds, bonnet motif, front wing flashes and rear number-plate lamp cover. For easy identification the Consul grille was oblong in shape with vertical bars, whilst that on the Zephyr Six featured a raised centre section and horizontal bars. A chrome/enamel badge — a lion rampant and three swords — was mounted on the central bar of the Consul's grille, and a 'winged' version of this was positioned just above the grille on the Zephyr Six and repeated on the boot lid at the rear. A small range of paint colours was offered, which included, in addition to the inevitable Black, Bristol Fawn, Edinburgh Green, Canterbury Green, Winchester Blue and Dorchester Grey. A very subtle pale blue, named Opal, was an additional colour for export only.

The overall measurements were a compact 13ft 6½in and 14ft 3¾in, respectively for the four- and six-cylinder models, with a width of 5ft 4in. The overall height was given as 5ft 0¾in in both cases, despite the difference in tyre sizes, although this difference was apparent in the listed ground clearances of 6½in and 7in. In relation to these dimensions the unladen weights of 21.77cwt and 23.1cwt indicated a particularly sturdy overall design.

Although production was still some weeks away, prototypes which represented the production specification were ready in time for the 1950 Motor Show at Earls Court, and on opening day, October 18, the Consul and Zephyr Six were unveiled. Billed as the 'Five Star Cars', the new models received tremendous acclaim from both the public and the motoring press and, indeed, in the opinion of many people, they were the stars at Earls Court that year. The stylists had produced a very well-balanced scale-down of the 'three-box' shape devised in Dearborn, and this, coupled with the monocoque construction, oversquare engines, MacPherson struts and pendant pedals, was a revolutionary combination in 1950, although 20 years later it would be the norm for medium/large quantity production saloons. So Dagenham had produced, probably quite unwittingly, by far the most significant postwar car that had yet appeared from a British manufacturer.

At a tax-paid price of £532, the Consul was competing against cars like the smaller 1.2-litre Austin Devon at £537 and 1¼-litre side-valve Hillman Minx at £543, and the similarly-sized 1½-litre side-valve Morris Oxford at £573. The Zephyr Six at £608 had to contend with the 2.1-litre four-cylinder Standard Vanguard at £703, the 2.2-litre four-cylinder Austin

The nearside view of a 1951 Zephyr Six.

Hereford at £738 and the 2.2-litre six-cylinder Morris Six at £718. Although similarly priced, at £531 and £601, Vauxhall's Wyvern and Velox were still hastily revamped prewar models, a fact evident in both their appearance and their performance, and so were not serious contenders for the time being.

After studying the specification and examining the cars at close quarters, the question uppermost in the minds of many of the 475,326 visitors to Earls Court was 'How can Ford do it at the price?'. The answer to this question lay in the completely new manufacturing and assembly techniques which had been devised at Dagenham hand in hand with the development of the new models. New machining processes that were in advance of anything used in the world before had been introduced. To illustrate this point it is necessary to explain that in machining a cylinder block, for example, the practice up to that time had been to break the process down into separate stages with a machine tool and operator at

The prewar styling of the bottom-of-the-range Anglia contrasts sharply with the Zephyr Six in 1951.

The lack of stowage space in the original flat dashboard is evident in this shot.

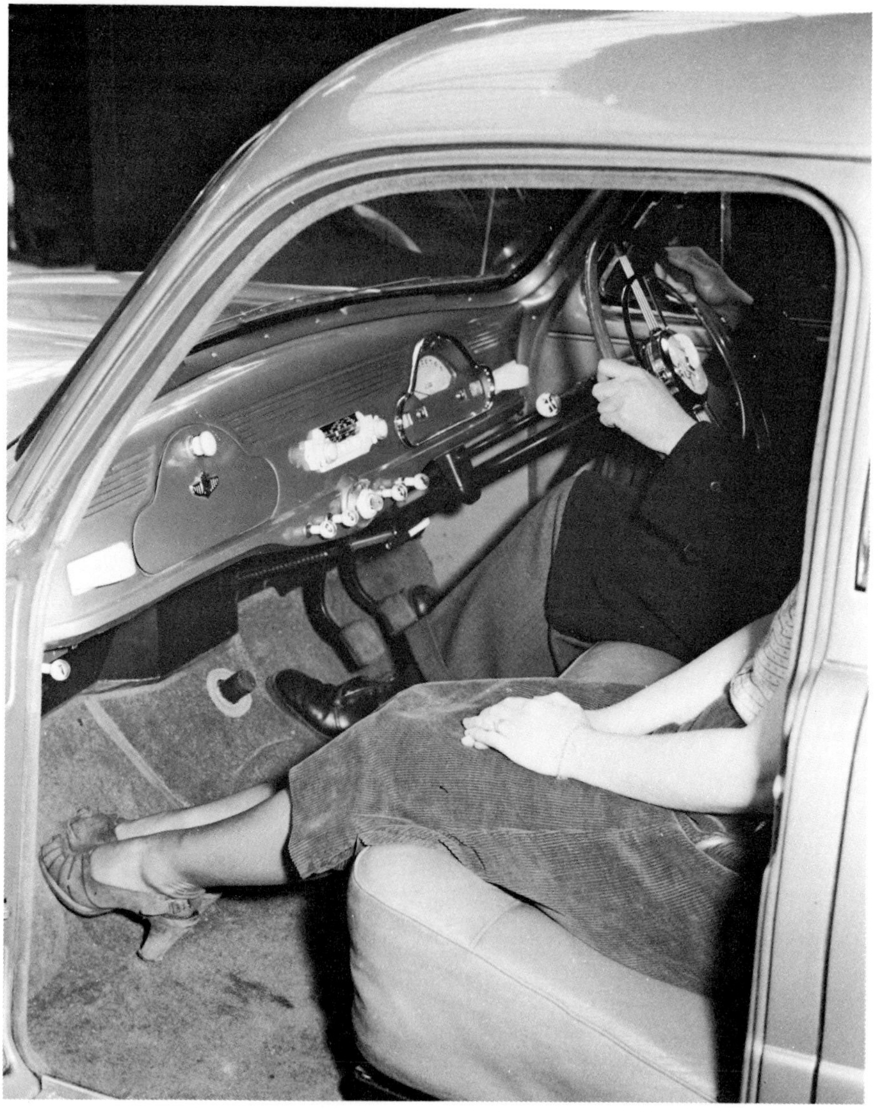

each stage. The cylinder block would have to be moved by hoist, or sometimes even manhandled, to the next machine tool after each operation. The new line of machines that was producing the Consul and Zephyr cylinder blocks at Dagenham was known as a 'transfer' or 'process through' line, and eliminated the handling of the block between operations by employing a series of mechanical/electrical/hydraulic sequences to move the block from one stage to the next. The separate machine tools on this new type of line were arranged in 'tunnel' formation on each side of a central rail on which the cylinder block was mounted. At the press of a button, the block would be moved on the rail to the first stage, where it would be automatically clamped in position for the machining operation. This pressbutton sequence would then be repeated for every stage and a fully machined cylinder block would emerge at the end of the line. This transfer line, which had a production rate approximately 100% greater than the earlier method, was considered at that time by the Ford Motor Company to be 'the greatest single advance in the production engineering

The optional radio was styled into the facia, with the speaker installed above the windscreen.

of motor vehicles since the industry commenced'. Similarly, crankshaft machining had previously been carried out in sequence on four different machines at Dagenham, but only one new machine tool was now required to complete the same four operations.

The final assembly line, too, was completely new, and once again far in advance of anything else in Britain at that time. Pit work had been eliminated, as this was felt to be particularly hard on the assembly line workers, and in order to allow the workers the fullest possible freedom of movement around the cars, the line itself had been sunk in to floor level wherever this was convenient.

Meanwhile, as the crowds at Earls Court were admiring the new cars on the Ford stand, some of the other hand-built prototypes were undergoing exhaustive 24-hours-a-day testing in Sweden, Belgium and the Swiss Alps, whilst others were taking advantage of the testing facilities of Ford France and the parent company in America. After a careful analysis and cross-checking of test reports had been completed, a production date was set, and on January 1, 1951, Ford of Britain Chairman Sir Rowland Smith drove the first Consul off the new assembly line.

Production of the four-cylinder car built up quickly over the next few weeks, and was joined on February 12 by the first Zephyr Six, although production of this version was to be in limited numbers only for some time to come, as the more economical Consul was rather more relevant in the

1951 Zephyr Six with whitewall tyres, and wheeltrim rings similar to those adopted later as a standard Zodiac feature.

still somewhat austere postwar world. In any case, in Britain, 1951 — 'Festival of Britain Year' — was to be a time of extreme frustration for the would-be new car buyer as, despite the fact that orders for new cars of all makes totalled more than a million at garages throughout the country, the Government were restricting the home market to an allocation of just 80,000 new cars that year. This was in order to force the manufacturers to export the bulk of their production and so earn valuable foreign currency for Britain. With this quota fixed, the Government then doubled the purchase tax on new cars to a massive $66\frac{2}{3}\%$ in the April budget, knowing full well that, with over a dozen people chasing every new car coming on to the market, even this level of taxation would not affect total sales at all. The new tax-paid prices were £662 and £759 for the Consul and Zephyr Six, which still represented quite a bargain for those who were fortunate enough to find themselves at the top of the waiting list. Many customers were placing their orders with several different garages in the faint hope that just one of these orders would bear fruit; if it did, they would then have to sign a two-year covenant on taking delivery, which prevented them from selling the car within that specified period. However, with the enforced shortage of new cars pushing used car prices up to considerably more than those listed for similar new models, there were many people quite willing to pay 'over the odds', and so in many cases much profiteering took place before a signature finally appeared on the covenant.

As soon as production was running smoothly, a Consul was made available to the motoring press, and on April 13, 1951 *The Autocar* published a full road test of this car, SHK 857. In almost every respect the Consul came in for praise from *The Autocar* test staff, who claimed: 'It is one of the outstanding cars produced since the war in the popular class

and has handling qualities that would be acceptable on a car of any price'. Of the power unit it was said that 'the engine is smooth and lively and satisfactorily quiet — full advantage has been taken of modern ideas in the design in regard to limitation of piston speed', and that 'the engine pulls very well on top gear. There is, however, everything against letting it do so beyond certain limits, in view of the fact that there is a useful middle ratio in the three-speed gearbox, engaged by a steering column lever which, because the linkage is simple, works extremely well with only finger pressure'. The performance included a genuine 75mph maximum allied to a 0-60mph time of 31.1 seconds, with the flexibility of the engine being underlined by a 20-40mph time of 13.1 seconds on top gear. Some

Despite a considerable impact, the damage to this Zephyr Six has been largely confined to the area of contact. Although some distortion of the lower part of the scuttle is evident, the windscreen pillars and the screen itself have not been affected.

Original underbonnet layout. For 1952 and later models, the coil was repositioned alongside the engine, and the starter solenoid was also removed from the bulkhead to a position on the nearside, adjacent to the battery.

criticism was levelled at the lack of a parcel shelf in the front, and also at the poor turning circle of 41ft. SHK 857 then passed into the hands of *The Motor* test staff, and their report was published in the issue of May 23. Once again, apart from the mediocre turning circle and lack of stowage space in the front compartment, it was virtually all praise. Probably as a result of more running in, the performance of SHK 857 showed a worthwhile improvement in the hands of *The Motor*, for whom it turned in a 0-60mph time of 27.2 seconds and covered the 20-40mph range in an excellent 11.1 seconds on top gear. The maximum speed was fractionally down, at a best of 74.4mph. A constant-speed fuel consumption of 36mpg at 40mph suggested that an overall 28-30mpg should be possible for the average owner. Summing up, *The Motor* stated that 'its good all round performance and quite exceptional handling qualities make it that rare vehicle — an everyman's car that is basically good by any standards'.

Whilst SHK 857 was being put through its paces in the hands of the motoring press, a fleet of 22 Consuls and 22 Zephyrs, straight from the production line, were embarking on an intensive testing programme similar to that which the prototypes had been forced to endure, in order to determine as quickly as possible whether any serious differences existed in the durability of the mass-production cars by comparison with the hand-built prototypes. This programme involved once again sending cars abroad in search of extreme climatic conditions. In Scandinavia the cars were operated in Arctic temperatures whilst, in contrast, an extract from a report submitted by the Ford Motor Company of Egypt reads: 'The cars were tested over a sand route from Alexandria to Khartoum, involving a large proportion of sandy or rocky roads, stretches of hard track and virgin terrain, heavy rain having washed away normal tracks. In some places no passenger cars had ever been before'. Allied to the tests being conducted overseas were others at proving grounds in the United Kingdom, where full use was made of facilities such as ripple-type road surfaces, *pavé* and dust tunnels. The early results were encouraging to such an extent that it was decided to continue the tests with these early production cars on an extended 'life' basis.

It was not long before the new Fords were earning foreign currency for Britain, and many of the early cars were shipped in 'knocked down' form for assembly in overseas plants. The Ford Motor Company of India were soon assembling Consuls alongside the already established Prefect, Thames trucks and Fordson Major tractors at their assembly plant in Bombay. Consuls for Malaya were assembled at the Ford plant in Singapore, whilst other Far Eastern countries were taking built-up cars from Dagenham, a batch of Consuls being included in a shipment for Japan for sale exclusively to United States service personnel and American business organizations that had been established there. In June, the first export Zephyrs, in the hands of Ford drivers, took the cross-channel ferry to Dunkirk and then motored across France, Belgium and into Germany. Joining the autobahn at Aachen, the Zephyrs headed for Cologne and a visit to the German Ford works, before turning north to their destinations in Sweden, Norway and Denmark. Soon after this, 'knocked down' Zephyrs were arriving in Bombay and Singapore. On the home market, the small number of Consuls being released were building up an excellent reputation from the start, the only regular criticism being that already referred to by *The Motor* and *The Autocar* — the lack of anywhere other than the small cubby hole in the facia to stow small items.

Meanwhile, unable to sample a Zephyr Six for themselves, many regular

Consul facia and controls, from 1952. The knob for the starter (pull to operate) is situated centrally under the parcels shelf, and can just be seen beneath the steering wheel boss. The knob on the instrument housing is for the wipers, the toggle switch beneath is for panel lights, and the control marked 'B' is for the heater blower.

The restyled facia on the Zephyr Six. The ignition switch and starter button changed places later as, in the arrangement here, the ignition lock number could be read from outside the car. The apparently haphazard placing of the controls and switchgear was in fact a very convenient arrangement, carefully chosen so as to avoid confusion.

readers of the weekly motoring magazines were eagerly awaiting a road test report. First to oblige was *The Motor,* with a full road test published in the issue of October 3, 1951. The test staff were very favourably impressed, particularly with the car's outstanding combination of performance and roadholding, commenting that it had 'a general roadworthiness and stability which virtually take it out of the family motoring class and put it into the sports car category'. This enthusiasm was tempered somewhat by the nose-heavy weight distribution, which 'results in a decisive break-away at the rear end if corners are taken too fast'. Nevertheless they felt that 'in the general run of motoring conditions . . . the car handles in such a way as to instill very great confidence'. The maximum speed of the car tested was a best one-way of 81.1mph and 70mph was reached from rest through the gears in 29.9 seconds. This, coupled with an ability to go from 20-40mph and from 40-60mph in 8.5 and 11.8 seconds, respectively, in top gear indicated that the Zephyr Six would have no problems at all dismissing slower-moving traffic. This high performance was accompanied by a creditable overall fuel consumption of 23.7mpg. 'Finally one must say', *The Motor* concluded, 'that it represents one of the finest examples of value for money to be found today not only amongst English but also world production.'

Dagenham were quick to answer the criticism levelled at the lack of interior stowage space and, just a year after the announcement, the Consuls and Zephyrs on the Ford stand at Earls Court in the autumn of 1951 displayed a completely redesigned facia. Beneath a new dash panel of 'roll top' appearance was a full-width parcels tray, 35in wide and over 10in deep, able to accommodate objects up to 4½in high, and installed at a slight angle, sloping towards the front of the car, to prevent spillage. Also part of the redesign was a shroud encasing both the steering column and gear-change linkage, with a new binnacle instrument cluster above it. The instruments themselves were similar to the earlier type, but now had a black background and light-coloured figures, those on the Zephyr being

The optional radio in position in the restyled facia. This prototype car still retains the early throttle pedal.

'three dimensional' in appearance. A central, drawer-type ashtray was incorporated in the new dash. The control knobs were now of ivory plastic with black letters, and the Zephyr Six acquired a pushbutton starter in place of the pull-out cable which was retained in the Consul. On both cars, the original organ-type accelerator pedal operating through a hole in the toeboard was replaced by a suspended pedal, bringing it into line with those of the clutch and brake. All four doors received new trim panels with vertical flutes in the design matching those moulded into the new dash panel. In addition to these cosmetic improvements were a new door pillar striker plate incorporating a nylon block, and a reshaped petrol filler neck designed for easier filling. Cars with this revised specification could be identified externally by a new grille badge, based on the previous design but enlarged slightly and including the proud words 'A Ford Product. Made in England'. Commenting upon these changes, the Ford Motor Company explained: 'The majority of these changes refer to minor items but they do reflect the fact that our claim to a policy that is one of continuous improvement, includes every item, however small, that makes for more pleasurable and efficient motoring'.

Also exhibited at Earls Court in 1951 was a prototype Zephyr Six Convertible de Ville, and VIP visitor Princess Margaret spent a long time on the Ford stand examining this exotic looking car, although not even Her Royal Highness would have been able to purchase one as little development work had taken place and production was still a considerable time away. As a result of increases in material costs the Ford Motor Company had recently announced an increase in the prices of the entire passenger car range, and the Consul and Zephyr Six were now listed at a tax paid total of £717 and £816, respectively.

Following the Motor Show, in the motoring press in November, the Zephyr Six was certainly 'car of the month'. Raymond Mays, of prewar racing fame, had taken delivery of one of the first Zephyrs registered in Britain, and writing in *The Autocar* issue of November 9, 1951 he gave a glowing account of its performance so far. Explaining that much of his previous motoring experience had been in 'high performance vehicles of the grand luxe class', Mays went on to say of the Zephyr: 'I have never known a "cheap" car which left me at such a loss for disparagement — or so monotonously fulsome in praise. Sincerely and soberly, I regard this Ford's advent as one of the most important events in mid-century motoring'. In five months Raymond Mays had covered 11,800 miles in the Zephyr including a 2,500-mile weekend trip to Milan for the Italian Grand Prix. This trip took him over the Simplon Pass, and of this Mays said that 'the Simplon proved an eye opener; I estimate that 75 per cent of the climb was made on top gear. On that ratio we repeatedly came down as low as 20mph, then pulled away without a vestige of labouring or fuss. Always in the past when driving cars with three-speed boxes I have sooner or later found myself hankering for that fourth gear; on the Zephyr, never. If four speeds were an optional extra chargeable at £2, I wouldn't specify them'. He went on to praise its 'freedom from tilt under the stress of deliberately violent cornering tactics, rapid pick-up on top gear, roominess and superb comfort'. The only trouble Mays had experienced so far was seepage of brake fluid past a defective hydraulic seal, and he also noted that 'a just perceptible vibration' would sometimes 'creep in — although it doesn't always do so — at around 65mph'. Summing up, Mays claimed: 'The Zephyr is a truly outstanding car in almost every respect. It corners like a racing car — better than some racing cars, in fact. Where road conditions permit, you can make maximum speed and cruising speed synonymous, and it takes this sort of medicine indefinitely'.

The following week the Zephyr came in for more praise when the respected Technical Editor of *The Motor*, Laurence Pomeroy, writing in the issue of November 14 about a trip to Europe in one of the new Fords, said: 'Surely the Zephyr can be thought of as the impoverished gentleman's Aston Martin'. After this it was the turn of *The Autocar* once again with a test report of the Zephyr in the issue of November 23. This particular car was complete with the new facia panel and *The Autocar* noted that 'it now provides a parcels tray of ample proportions'. The performance of this very new Zephyr included impressive maxima in the three gears of 31mph, 56mph and 81mph, with 70mph reached in 32.1 seconds from rest. The steering and handling came in for the usual praise, with the comment: 'The steering is quite light and very positive, with good self-centering action. The slight yet definite amount of inherent understeer quickly inspires confidence and, in fact, the Zephyr steers extremely well on both fast main road bends and narrow winding roads'. Of the column-mounted gear-change it was said: 'The lever is much shorter than usual and very positive in operation; it is in some ways reminiscent of the remote control unit fitted on some sports cars'. *The Autocar* concluded: 'Judged on both performance and on value for money, the Zephyr is a very satisfactory car'.

By the end of 1951, exactly one year after the first Consul left the line, Dagenham had produced 35,667 examples of the four-cylinder car, 15,880 of which were in 'knocked down' form. The corresponding figure of 3,463 (950 'knocked down') for the Zephyr Six underlined its exclusivity so far, although a big increase was planned for 1952. The changes exhibited at Earls Court in October 1951 had only appeared on the development cars, and were not at that time ready for production, but they were to be phased in gradually during 1952. This resulted in certain cars displaying some, but not all, of the new features, and it was not until late in the year that cars exhibiting the fully updated specification were leaving the production line. In the meantime, for the benefit of owners of cars with the early flat dashboard, the Central Motor Company, of Leicester, were offering a neat accessory parcel shelf, priced at £2.

In April, Ken Wharton had driven his Consul to outright victory in the Dutch International Tulip Rally, but with long waiting lists already for this model at all Ford agencies, coupled with the planned increase in Zephyr production, there was little that Dagenham could do to capitalize on this particular success. Nevertheless, the victory added considerably to the Consul's prestige. Production continued to build up steadily throughout the year and, as the Zephyr Six was making its debut in America, the Consul was introduced in Australia, where it was to be assembled at the Ford works in Geelong, Victoria. From Geelong, the Consul would be distributed throughout Australia and Tasmania, whilst cars for New Zealand would be assembled by the Ford Motor Company of New Zealand Limited, at their plant in Lower Hutt.

The new facia panel had met with widespread approval, and so the basic design was now established for some considerable time ahead. No visible changes were planned for the 1953 model year, but a technical change was introduced in November 1952. The rear axle ratio on both models was altered, that on the Consul becoming 4.556:1 (4.625:1 previously). This raised the gearing slightly, giving a worthwhile improvement in the constant-speed fuel consumption in top gear, with a

Zephyr Six showing the repositioned badge for 1952 and later models.

figure of around 32-33mpg at 50mph. In conjunction with this change, slightly lower indirect ratios were introduced in the gearbox, actually lowering first and second gear overall, so giving improved acceleration from rest, but, on the debit side, opening up the gap between second and top. On the Zephyr, the axle ratio was now 4.444:1 (4.375:1 previously) and, as the Zephyr also received the new indirect ratios in the gearbox, its already remarkable accelerative powers were sharpened up even more. The speed at 5,000rpm in top gear was now 80.5mph, compared with 81.75mph with the earlier axle ratio, but as the engine would pull a few more revs on the slightly lower gearing the maximum speed was unchanged. At 5,000rpm on the Consul the theoretical speed would now be 76.5mph, an improvement of 2mph, although in practice the four-cylinder would not quite pull 5,000 revs on the higher gearing, so once again the actual maximum speed proved to be about the same as before.

On January 15, 1953, Dagenham's 2,000,000th vehicle was driven off the production line by Ford Chairman Sir Rowland Smith. The car, a left-hand-drive Zephyr Six, complete with whitewall tyres, was destined for the showrooms of Ford Dealers Holmes Tuttle and Company, of Los Angeles, California. An average of 400 Zephyrs a month, out of a total of 550 cars of

An early Zephyr estate car project, the 'Horsham', which did not reach production.

all models, was now being shipped to the United States from Dagenham. On the morning of January 27 the Zephyr Six was front page news; Dutchman Maurice Gatsonides had driven a standard Zephyr to outright victory in the Monte Carlo Rally. The following day, all the National dailies were carrying Zephyr adverts and, thanks to an overnight shift at a Dagenham printing works, every Zephyr leaving the line on January 28 was sporting a windscreen sticker proudly proclaiming the Monte Carlo win, whilst other batches of stickers were being rushed to dockside car parks wherever Zephyrs were awaiting shipment abroad. The winning

Zephyr was flown to London on January 30 and put on display in the Company's West End showrooms at 88 Regent Street.

During April 1953 the revised first and second gear ratios which had been introduced six months earlier were deleted in favour once again of those of the original specification. The revised *axle* ratios, however, were retained, and this new arrangement was standardized with no further modifications. So now, by comparison with the original 1951 production specification, all three gears on the Consul were raised, whilst all three on the Zephyr were very slightly lower than before, although the Zephyr was, of course, still the higher geared overall of the two cars.

In Australia the Consul had now been joined by the Zephyr Six, and the design team at Geelong, who in the past had produced variations on many Dagenham themes to suit local circumstances, were keen to produce a

One of the rare Zephyr Six utilities which Ford Australia had hoped to mass produce.

pick-up truck, or coupe utility as these are referred to in Australia, based on the Consul and Zephyr. Prototype utilities of both four- and six-cylinder models were produced, but, surprisingly, despite the enthusiasm of the Geelong team for this new version, Dagenham were not happy about building pick-ups based on a monocoque shell and eventually they vetoed the idea, so the design was shelved after only six Zephyr and five Consul utilities had been produced.

After the Monte Carlo success, many enthusiasts who had hitherto only looked upon the Zephyr as a family car were now taking another look. For the benefit of *Autosport* readers, John Bolster was busy putting a Zephyr through its paces, and his report appeared in the issue of May 8. Starting with a brief technical description of the car, he went on: 'I first took over the Zephyr in London, and I must admit that it did not endear itself to me immediately. Every time I braked at low speeds, the nose dipped considerably, and it rose sharply into the air on sudden acceleration. All this gave me the impression that, for the next few days, I was going to be stuck with a sick-making, American-style car. How wrong I was! Once out on the open road, the machine's real character asserted itself. It seemed to like cruising at almost its maximum speed, at which velocity a very level ride is provided'. Of the handling he commented: 'Perhaps the most surprising feature is the lack of serious roll on corners — the body remains more vertical during vigorous manoeuvres than is common among present day cars. The steering is light and direct, and it is "quick" enough to deal with emergencies. On dry roads, the car corners well; on wet and slippery surfaces, however, one becomes conscious that the rear wheels are relatively lightly loaded — wheelspin occurs if one accelerates violently in first or second gear, but the resulting tail slides are always easily controllable'. About the three-speed gearbox, John Bolster echoed Raymond Mays' sentiments, with the comment that 'this is one of the few three-speeders which would not be improved by a fourth ratio'. After suggesting that its light controls, good top-gear acceleration and good visibility 'makes the Zephyr an ideal woman's car', he went on: 'From my own point of view, I found that this machine invited me to drive it disgracefully hard, and I was nothing loath to oblige'. Summing up, he wrote: 'The man who buys a Zephyr gets an astonishing amount of performance and comfort for his money. It is, first and foremost, an everyday family car, but under its bonnet there is enough "zip" to make it an agreeable companion for the chap who is always in a hurry'. Checked against the stopwatch, the Zephyr had recorded a two-way average maximum of 80mph and reached 50, 60 and 70mph from rest in 12.8, 19.6 and 29.6 seconds, respectively, whilst the standing-start quarter-mile had occupied exactly 21 seconds.

Amplifying the comments by John Bolster about the wheelspin in wet conditions, it was certainly considered in some quarters that the Zephyr could be quite a handful in the wet. The vast majority of Zephyr buyers were sampling high-performance six-cylinder motoring for the first time, virtually all their previous experience having been on prewar or prewar-inspired cars, many with side-valve engines which were so lacking in throttle response that the driver was able to floor the accelerator pedal and then, quite literally, wait for signs of progress. In contrast, the Zephyr engine responded immediately to any change in the position of the pedal, and many drivers had simply never learned a progressive, light throttle technique, largely because it had not been necessary on anything they had driven before. If the Zephyr had been selling at twice the price, it is almost certain that it would never have earned this 'tail happy' reputation, as it would only have been bought by those drivers who largely were already used to powerful cars and had therefore developed the required driving technique. John Bolster's observation that 'the resulting tail slides are always easily controllable' was absolutely correct, but here was speaking a man with a vast experience of very high-performance machinery, including racing cars, to whom controlling a tail slide was, in any case, second nature. To many ordinary motorists, though, a tail slide, or 'skid' as most of them called it, was something to be feared. Writing about the same subject in the June 24, 1953 issue of *The Motor,* Technical Editor Laurence Pomeroy accurately summed up the situation with his comment that it was possible to 'promote ready wheelspin and a quick breakaway at the back. But due to the direct and positive steering such slides can be quickly ended, or accurately controlled, according to choice, and although the Zephyr can present problems to the novice, it is a safe and pleasing car for the experienced man'.

Laurence Pomeroy had commented favourably about the Zephyr on several occasions in *The Motor,* and had recently had some interesting modifications carried out on the engine of his own example. The compression ratio was raised to approximately 7.5:1 and the inlet and exhaust ports smoothed out. Double valve springs were fitted to cope with the anticipated extra rpm, and triple SU carburettors on separate induction pipes replaced the single Zenith. A large cylindrical Vokes air filter/silencer with a curved three-branch pipe leading to the carburettor intakes was mounted on the rocker cover. The exhaust arrangements consisted of two separate systems, each comprising a three-branch manifold feeding into its own silencer and tailpipe. The result of these modifications was a maximum speed of over 95mph coupled with a substantial increase in acceleration. The modifications on this particular Zephyr were very definitely on a 'one-off' basis, but elsewhere plans were in hand to market a similar conversion.

Zephyr enthusiast Raymond Mays had also been quick to realize the tuning potential of the short-stroke engine, and for some time now his Zephyr had been running in a remarkably similar state of tune to that which Laurence Pomeroy had devised. Triple SU carburettors, improved ports, double valve springs and a 7.5:1 compression ratio were all features of the Mays Zephyr. A single exhaust system was retained in this instance, although the original baffle-type silencer was replaced with one of the Servais straight-through variety. After 30,000 hard miles in this state of tune the Zephyr engine was showing no ill effects whatsoever, and Mays decided to offer this conversion for sale to the public at an inclusive price of £75.

John Bolster borrowed Raymond Mays' own Zephyr for a weekend's quick motoring, and reported enthusiastically on the experience in the *Autosport* issue of September 11, 1953. 'On taking the wheel', he wrote, 'it was at once obvious that the car had a lot more "steam" than any Zephyr I had previously driven. A full 30mph came up like a flash on bottom gear and, rather pointlessly, I was guilty of touching 60mph on second speed. On top, that "out of breath" feeling at 80mph had quite gone, and the acceleration continued in one smooth surge. To give an idea of the increase in performance I might mention that the modified car accelerated from a standstill to 80mph in the same time that the standard job took for the 0-70mph range. The maximum speed was over 90mph and at that velocity the engine was as smooth as a little dynamo. Curiously enough,

An early Zephyr Six convertible. Although right-hand-drive, the two-tone paint and whitewall tyres are thought to be the American export finish.

the improved high-speed behaviour of the unit made one much less conscious of the need for an overdrive. To maintain such a rate involves some pretty impressive rpm, but the short-stroke engine makes no complaint'. Summing up his impressions of the Mays Zephyr he wrote: 'After several hundred miles of exceptionally quick motoring, I came to the conclusion that Messrs Raymond Mays and Partners, of Bourne, Lincs, know just about all the answers where the Zephyr Six is concerned. It only remains to add that the success of the conversion is a fine advertisement for the basic soundness of Ford design and construction'.

With the easing of home market restrictions, more and more Consuls and Zephyrs were to be seen on the roads of Britain as Dagenham's 'Five Star Cars' firmly established themselves as the number one sellers in their particular sector of the market; and now the saloons were at last being joined by the long-awaited Convertible de Ville. Carbodies of Coventry had been building drophead coupes since the early 1920s and were first approached by Ford soon after postwar production had commenced at Dagenham, although the result was only a small number of prototype Anglia and Prefect convertibles, the projected Consul/Zephyr being a more attractive proposition in the long term. Development of the Zephyr convertible had taken much longer than had been anticipated, however, as, deprived of the steel roof, which played an important part in the strength of the saloon body, the floorpan was under too much strain and, on test, prototypes had actually suffered with cracking floors. Although it was generally accepted that convertible bodies were always appreciably less rigid than their saloon counterparts, in view of the performance available with the Zephyr it was important to reduce this difference to a minimum before production could commence, and extra bracing, in the form of a substantial X-frame, was developed by Carbodies and welded into the standard floorpan supplied by Dagenham.

The hood was in a very durable, washable, plastic fabric, and was

When in the raised position, the weatherproof hood gave a slightly lower look to the convertible models.

An early Consul convertible. The bonnet motif is non-standard and, although similar to that of the early Zephyr Six, is a Wilmot Breeden accessory in this case, which was available for £1.5s.

complete with a safety glass rear window. To and from the de Ville (halfway) position, the hood was hydraulically operated from a switch on the dashboard, and a particularly neat touch was that the rear seat tilted forward slightly as the hood was being raised or lowered, ensuring that the folds in the top did not disturb the passengers' heads. The backrest of the front bench seat was divided, with each side hinged to allow easy access to the rear compartment through the wide and rather heavy doors. All four side windows could be raised or lowered irrespective of the position of the hood, giving a remarkable overall versatility, and the only obvious shortcoming of the car was the lack of sun visors and an interior light, particularly surprising as there was ample room for both these features on the windscreen surround. At £960 the Zephyr convertible was £206 more than the saloon, but still over £300 cheaper than the nearest comparable

The exterior sun visor was a popular 1950s extra.

drophead coupe, the $2\frac{1}{4}$-litre Sunbeam Talbot 90 convertible at £1,269.

A Consul convertible, with manually-operated hood, followed quickly: at £808 it had to contend with the Austin Somerset coupe, which was very attractively priced at £705, and the rather smaller side-valve-engined Hillman Minx drophead at £723, so it was not really as well placed in its class as the virtually unrivalled Zephyr Six. Most of the early Zephyr convertibles were destined for overseas markets where, in addition to the standard range of colours, they were also available in Mandarin Red with silver grey upholstery, or Ivory with ivory. Whitewhall tyres completed the export specification.

There had not been a 'de luxe' version of the Consul or Zephyr, although a heater and leather upholstery had been listed as options right from the start. Of course owners could, and indeed often did, purchase spot lamps, wing mirrors and so on from their local garage or accessory shop. In view of the popularity of these extras, the Product Planning executives at Dagenham had decided early in 1953 that the time was ripe to add a comprehensively equipped model to the top of the Consul/Zephyr range, the specification of which would go considerably further than the typical 'de luxe' offerings of rival manufacturers. The result of this decision, the Zephyr Zodiac, was introduced to the public on October 21, 1953, the opening day of the Earls Court Motor Show, and was, in the opinion of many people, the most glamorous car yet to appear from a British manufacturer.

The new car was instantly recognizable externally in a range of beautiful dual-tone colour schemes, the division of which was highlighted by a new front wing chrome flash, incorporating the 'Five Star' motif, and a chromium-plated moulding running the length of the body. New deeper-section bumpers and overriders were introduced, as was a futuristically-styled 'aeroplane' bonnet mascot to replace the original rocket-shaped fitting. Spot, fog and reversing lamps, wing mirrors with spring-loaded stems, chromed lockable fuel-filler cap, chromed wheelrim embellishers, whitewall tyres and 9-carat gold-plated name scripts completed the exterior scene. The passenger compartment featured leather upholstery

A prototype Zephyr Zodiac. The script on the bonnet front still reads Zephyr Six, though it is placed higher up on the bonnet so as not to be obscured by fog lamps; the original holes for the fixing can be seen in the lower position.

and restyled door trims, both of which were finished in a two-tone colour scheme complementing that of the exterior paintwork. A new woollen headlining replaced the cloth lining used in the Consul and Zephyr, and a larger rubber heel pad than that of the Zephyr was incorporated in the carpet in front of the driver's seat. A cigar lighter, a vanity mirror in the passenger's sun visor and an electric clock were other standard features,

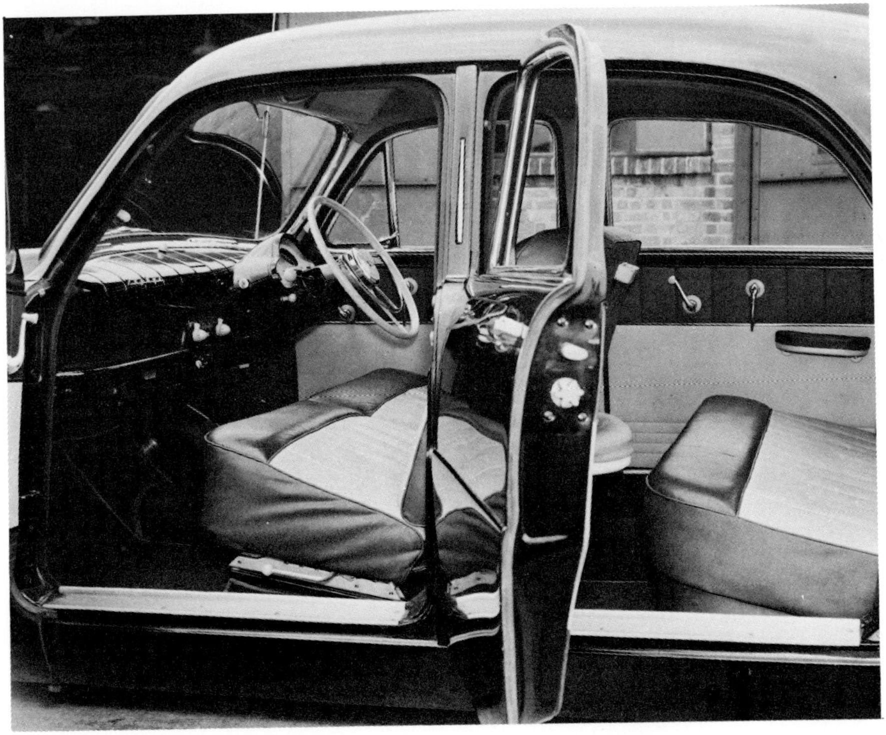

1953 Zephyr Six with non-standard interior trim, including separate front seats. This is presumably a Styling Department car, and would appear to have been involved in the Zodiac development programme.

Zephyr Zodiac, regarded by many Ford enthusiasts today as one of Dagenham's finest-ever cars.

and the switch in the steering column boss now operated flashing direction indicators incorporated in the front and rear side lamp units in place of the original semaphore type. For additional safety, the rear light surround also incorporated a circular reflector.

Under the bonnet was a large, heavy-duty battery, a vacuum-operated windscreen washer system and the previously optional heater and demister unit. The engine boasted a new cylinder head of 7.5:1 compression ratio, raising the power output from the standard Zephyr's 68bhp at 4,000rpm to 71bhp at 4,200rpm, and being designed to make full use of the recently-introduced premium grades of petrol, this cylinder head and its rocker cover being painted bright red. In the luggage compartment was a padded steel toolbox, placed between the spare wheel and the inner wing, and complete with two box spanners, two double-ended spanners, an adjustable spanner, a plug spanner, two screwdrivers, a tommy bar, a tyre lever and a brake adjuster, these being in addition to the grease gun, wheelbrace and jack, which were normally supplied with the Consul and Zephyr.

In addition to the three dual-tone colour schemes of Dorchester Grey upper half combined with Canterbury Green, Winchester Blue, or Bristol Fawn lower half, the Zephyr Zodiac was also available, to special order only, in single-tone black, and in this finish it came with the Dorchester Grey/Bristol Fawn interior trim combination. At a tax paid price of £851,

Zephyr Six interior, from October 1953 onwards. Zephyr Zodiac interior trim was similar in style, but with a two-tone finish to complement the paintwork.

the new model had no obvious rivals in its own price range and was, according to the Ford Motor Company, 'destined to take its place among the products of a select and exclusive band of British manufacturers who have set standards of quality workmanship that have become the envy of the world'. Surprisingly, a Zodiac convertible was not offered, although just one was subsequently made for Mrs Benson Ford, wife of Henry Ford's second grandson.

The new interior trim styling of the Zephyr Zodiac was now applied to the Zephyr Six, although still in single-tone PVC, with leather available at extra cost. The Zephyr Six also acquired the 'Five Star' flashes and side strips, the new bumpers and overriders, the aeroplane-style mascot and the flashing direction indicators, whilst its cloth roof lining was now of a tighter, more dust-resistant weave. The Consul received the new bumpers, but still without the overriders, which were to remain an extra-cost option on the four-cylinder car. It also received the new side strips and its own variation of the 'Five Star' wing flashes, a simple bonnet mascot and, at last, chrome inserts in the front and rear screen rubbers. Inside, the Consul now featured an improved rubber floor covering, colour keyed to the rest of the interior, and a restyled pattern on the seats and door trims, whilst under the bonnet a single windtone horn replaced the rather ineffective high-frequency unit fitted previously. Some technical changes, to improve efficiency and all-round performance rather than to increase showroom appeal, had recently been phased into production, and included improved clutch drive plates with six damper springs instead of the previous four or five springs in the Consul and Zephyr, respectively, and chromium-plated top piston rings, which would further improve the hard wearing qualities

The position of the electric clock in the Zephyr Zodiac. The interior light is manually operated.

that the Consul/Zephyr engines already displayed. Once production of the Zodiac was under way, the 7.5:1 compression ratio cylinder head was made available to buyers of the Zephyr Six as a no-cost option, retaining its red paint finish for identification.

The whole range had now been very effectively extended and updated, without having to resort to expensive retooling costs, as the body panels were exactly the same as before, and with the tax paid prices of the Consul and Zephyr Six remaining unaltered at £666 and £754, sales continued to expand.

In production, the new two-tone Zodiac bodies were first painted all over in the upper-half colour of Dorchester Grey, and this was stoved in the normal manner in a conventional convector-heated oven. The lower half colour was then sprayed on over the grey, and the body then passed into another oven with infra-red heaters adjusted exactly so as to stove the bottom half only, as restoving the top could have caused discolouration.

Consul with front wing chrome trim, a style which was not adopted for production.

Consul, as produced from October 1953, with optional overriders.

The bonnet is first released by pulling a T-handle under the facia, and then by pressing the safety catch as shown here.

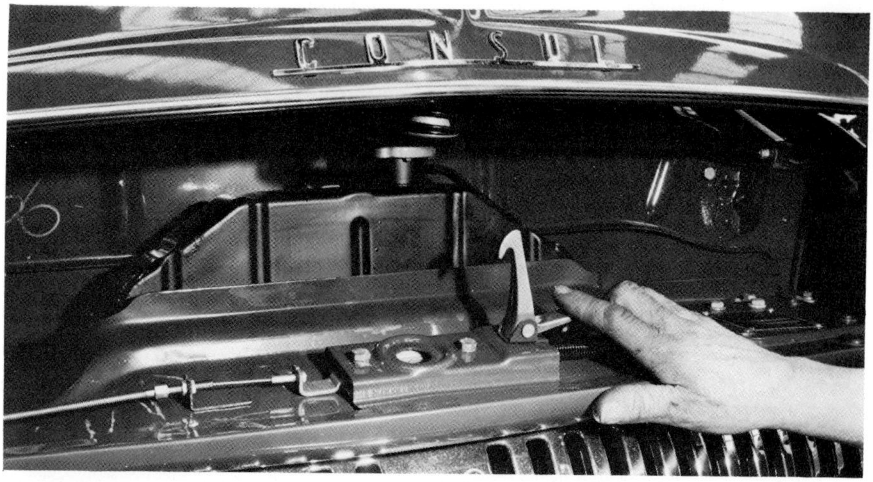

The new Prefect, introduced in 1953, displayed a marked similarity to the Consul, and inherited the MacPherson front suspension arrangement of the Consul/Zephyr range.

This localized infra-red oven had been designed specially for the Zodiac project by Briggs Motor Bodies in conjunction with a team of engineers from Metropolitan-Vickers Electrical Company.

In the issue of July 16, 1954, *The Autocar* published a full road test of the Zephyr Zodiac. With a price tag of £851 allied to such an impressive specification, the fittings and furnishings were beyond criticism, and even though the basic design was now four years old, the performance and handling came in for the usual praise, the test staff commenting: 'The engine is extremely flexible and pulls strongly from below 12mph in top gear up to its very creditable maximum without the slightest sign of roughness or vibration. When required, the majority of main road hills can be stormed in a very heartening fashion. . . . In addition to the excellent acceleration, it will cruise without effort at a genuine 70mph when carrying three large passengers and a full complement of luggage'. Of the gear-change *The Autocar* said: 'When it is necessary to change gear, the short, stiff, steering column-mounted lever makes the operation a pleasure', and of the ride and handling, 'the Zephyr Zodiac gives the driver and passengers an excellent ride in all circumstances. The ride in the back seat is very good and it is quite possible for two people to fall asleep in the rear compartment when the car is cruising at quite high speeds . . . the steering is positive and light . . . and the car feels as safe on wet main roads at speed as

Fixed-price routine servicing contributed to the remarkable overall economy of Consul or Zephyr ownership. On July 7, 1953 Ford had announced that new or fully reconditioned engines were available for the Consul, at £31, and the Zephyr Six, at £42, on an exchange basis.

it does on winding, badly surfaced by-roads'. The car tested reached maximum speeds in the three gears of 31, 57 and 84mph, and accelerated from rest to 30, 50 and 70mph in 5.4, 13.5 and 29.7 seconds, this performance being accompanied by an overall fuel consumption of 23.7mpg.

Many specialist tuning firms had followed the lead set by Raymond Mays and introduced conversion kits for both the Consul and Zephyr, which gave the enthusiast various stages of improved performance from which to choose. The Aquaplane Company, of Suffolk, were offering a redesigned inlet manifold with either twin (for the Consul) or triple (for the Zephyr/ Zodiac) $1\frac{1}{2}$in SU carburettors, complete with air filters, priced at £28 15s and £38 15s, respectively. These twin and triple set-ups could be relied upon to produce a worthwhile increase in acceleration throughout the range, particularly so at the top end, in conjunction with a slightly higher top speed, and were also attractive because of their 'bolt-on' nature, not requiring any dismantling of the engine. For the Zephyr and Zodiac models, Sports Autos, of Bayswater, could supply two stages of tune, the first of which comprised a new inlet manifold with two additional standard Zenith carburettors, the original Zenith being utilized to complete the triple set-up whilst keeping the price down to an attractive £29 10s. Stage 2, for an extra £9, involved removing the cylinder head, polishing the combustion chambers, easing the valve ports and fitting double valve springs. Sports Autos did not quote any performance figures in their advertisements, but it would be reasonable to assume that in Stage 2 form a Zephyr would be a 90mph car with a substantial improvement in acceleration. V.W. Derrington, of Kingston-upon-Thames, offered a two-stage improvement for the Consul, Stage 1 of which consisted of a twin SU

1954 Zephyr Six convertible. This particular car was a press demonstrator, and was road-tested by The Motor.

arrangement and a four-branch exhaust manifold, giving a claimed 3-second improvement in the 0-50mph time and a maximum speed of 84mph for a price of £30. Another £20 would buy Stage 2, which was the addition of a reworked cylinder head and would, according to the advertisement, knock another 2 seconds off the 0-50mph time, bringing it down to 12 seconds.

Laystall Engineering experimented with both Consul and Zephyr engines, eventually standardizing a comprehensive one-stage conversion kit for both models, priced at £45 and £65, respectively, with an additional £5 fitting charge, the customer's original cylinder head being taken in exchange. The compression ratio was raised to 8.3:1, and larger valves

A 1954 Zephyr Six displays its roomy luggage compartment, although on this occasion some squashing would be necessary to close the lid.

Tail-light surround for the Consul/Zephyr/Zodiac from October 1954 onwards.

were accommodated in opened-out ports and controlled by double valve springs. Twin or triple SU carburettors (Consul or Zephyr) supplied the mixture. At 4,400rpm, where the standard Consul engine was producing 47bhp, the Laystall unit was already showing 64bhp, and this continued to rise to 66.8bhp at 5,200rpm, at which revs the standard output had fallen to about 40bhp. The corresponding figures for the six-cylinder engine were: standard unit 68bhp at 4,000rpm; Laystall 87bhp, rising to 96.5 at 5,200rpm, by which time the standard unit was down to 50bhp. These figures indicated that a lot of performance was on offer, and Laystall invited John Bolster to sample both models for the benefit of *Autosport* readers. The Consul was the first available, and proved capable of a rate of acceleration through the gears that was not only an improvement over the standard model, but also over the standard Zephyr Six; from rest to 50, 60 and 70mph was covered in a remarkable 11.2, 16.8 and 23.6 seconds,respectively. A maximum speed of 84mph was recorded and the car returned an overall consumption of 23.6mpg. The normal flexibility was apparently in no way impaired, as John Bolster stated: 'The Laystall modified Consul can show a clean pair of heels to many small sports models, and yet it is in no way inferior to the standard job for shopping and business transport'. As can no doubt be imagined, the improvement with the modified Zephyr was considerable, and the figures recorded by John Bolster contrast dramatically with those for the standard Zephyr Zodiac tested by *The Autocar.*

The Autocar: Standard Zephyr Zodiac		*Autosport:* Laystall Zephyr Six
Acceleration times (seconds):		
0-30mph	5.4	3.2
0-50mph	13.5	8.8
0-60mph	20.4	13.0
0-70mph	29.7	18.4
0-80mph	—	25.2
0-90mph	—	37.4
S/S $\frac{1}{4}$-mile	21.5	19.2
Maximum speed	84mph	98mph

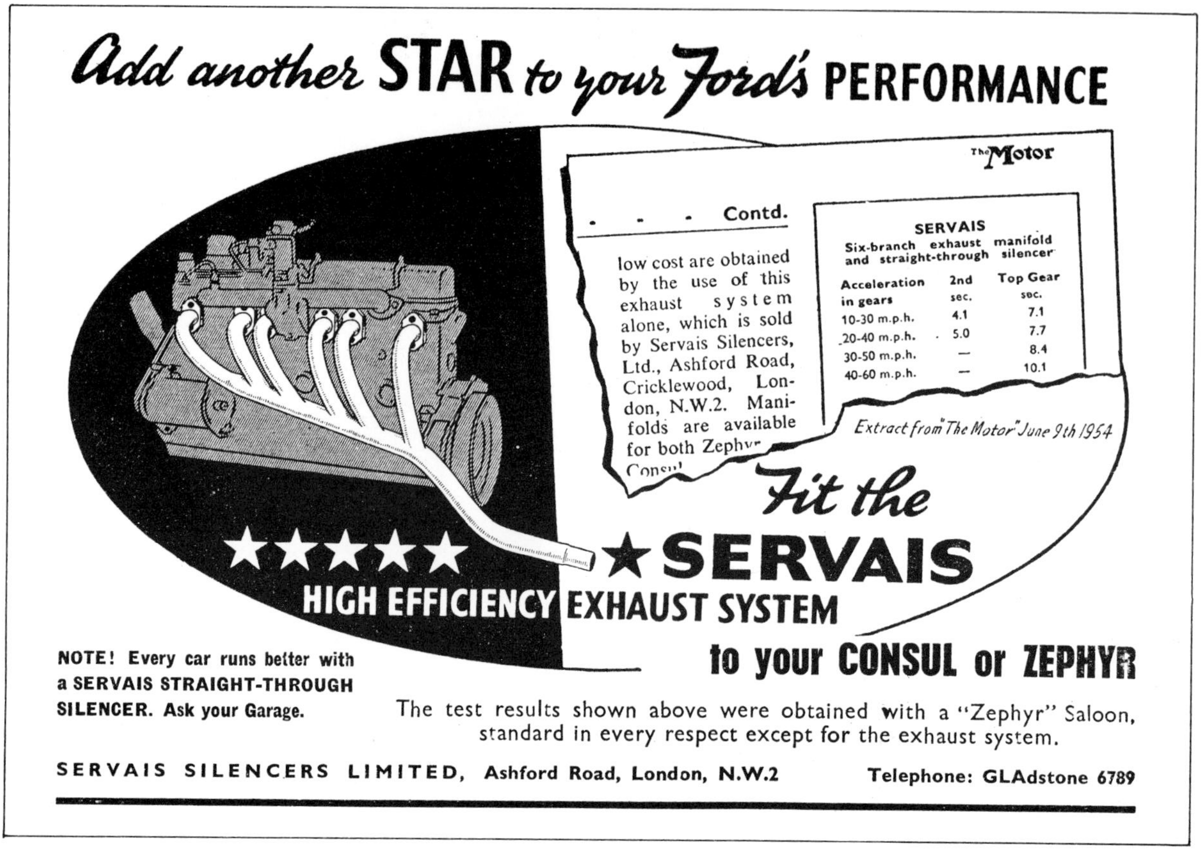

A Servais advertisement in which the illustration makes an interesting comparison with the earlier view of the standard Ford exhaust manifold.

For those motorists who wished for something in reserve over the standard model, these conversions were certainly well worthwhile. But it would have been unreasonable, for instance, to expect to habitually use the performance of the modified Consul to the full without some extra wear and tear taking place in the long term, and in this respect the hard-driving high-mileage driver would still have been better advised to go for the standard Zephyr rather than the modified Consul. Nevertheless, these figures do show that a considerable safety margin was built in to the standard engines, and the modified Zephyr in particular was offering outstanding performance in relation to the total price.

The standard exhaust manifold did apparently cause considerable back pressure, and Servais Silencers, of Cricklewood, developed a four- or six-branch manifold for the Consul or Zephyr designed to use their 'straight through' silencer and the original Ford tailpipe. Servais advertisements quoted the following figures for the Zephyr Six with this system; recorded by *The Motor*, they are top gear acceleration times in seconds, with the figures for the standard car in brackets. 20-40mph: 7.7 (8.2); 30-50mph: 8.4 (9.4); and 40-60mph: 10.1 (11.8). Servais developed this system a stage further, and later were able to offer a complete dual system for the Zephyr/Zodiac, priced at £26 5s and featuring a pair of three-branch manifolds, dual silencer and twin tailpipes. This gave acceleration times in top gear from 20-40, 30-50, 40-60 and 50-70mph of 7.2, 7.3, 9.1 and 11.5 seconds, respectively, which represented about the best that could be achieved with exhaust system modifications only. Another firm, at first calling themselves Consul Productions, but soon changing to Consela

E. D. Abbott's estate conversion retained many of the saloon's features, and utilized the original rear window.

Productions, advertised four- and six-branch manifolds at £9 5s and £12 10s, designed for use with the original Ford silencer and tailpipe.

At Earls Court in 1954 the few changes to be seen on the Consul/Zephyr/Zodiac range had in fact first been exhibited three weeks earlier at the Paris Show. All three cars featured a new rear light surround incorporating a reflector above the lamp unit. The rear flashing indicators were improved by a device which cancelled the appropriate stop light when either indicator was flashing. Inside, the parcel shelf had an improved covering, and the front seat adjustment range was increased slightly. The Zephyr Six acquired the larger rubber heel pad for the driver that was already a Zodiac feature, and the Zodiac's woollen headlining gave way to one of washable embossed vinyl. A clutch pedal assist spring

The rather high sill made loading somewhat more difficult than with some purpose-built estate cars.

Definitely not a 'van with windows', this conversion showed that an estate car could combine style and practicality.

was introduced on both the six-cylinder models in answer to recent criticism that the clutch was rather heavy to operate.

The most interesting Ford exhibit, however, was not on the Ford stand, but on that of the coachbuilders E.D. Abbott, of Farnham in Surrey. Abbotts had carried out an estate car conversion on a Zephyr saloon. The saloon rear quarter roof panels were partially cut away, and new rear side panels were welded in along the top of the wings. The roof extension was lower than the saloon roof, giving a slightly stepped effect, a deliberate feature to utilize as much as possible of the strength of the original domed roof. The extended roof section was complete with rubber-covered slats, a chromed roof rail and two strap eyes. A side-opening tailgate was employed, fitted with the original saloon rear window. By retaining the normal rear passenger doors much of the saloon's characteristic appearance had been kept, without a 'van with windows' look. Abbotts could carry out this conversion on any existing Consul/Zephyr/Zodiac saloon at an inclusive price of £145, and as the conversion met with the approval of the Ford Motor Company, arrangements could be made to have a partially completed new car converted by Abbotts, although purchase tax of £60 was then payable on the conversion.

In their issue of November 24, 1954 *The Motor* published a road test report on the Zephyr Six Convertible. The performance and handling was, of course, similar to the saloon's and were such, according to *The Motor*, as to have 'led many people to accept this model as a good substitute for a sports car'. The test staff noticed that 'at times slight shake suggests that the convertible body is a little less rigid than that of the all-steel saloon', although 'actual rattles were almost entirely absent'.

Elsewhere, another very sporting Zephyr was on test. Raymond Mays had decided to offer a genuine 100mph conversion. The weight and frontal area of the Zephyr, coupled with its relatively low overall gearing, were such that over 100bhp would be needed to reach a true 100mph, and this figure was difficult to achieve by modifying a standard cylinder head because of the restrictions imposed by the siamesed inlet ports. Mays therefore designed a new light-alloy cylinder head of 8.2:1 compression ratio, with six separate, steeply inclined inlet ports. The standard valves were retained, and the efficiency of the six-port arrangement was such that twin SU carburettors were now sufficient for the desired power output, and 106bhp at 5,000rpm was the result of these modifications. A speed of 100mph was now available, but at that rate the engine was turning over at 6,200rpm on the normal top gear ratio. Although not a Ford option, the

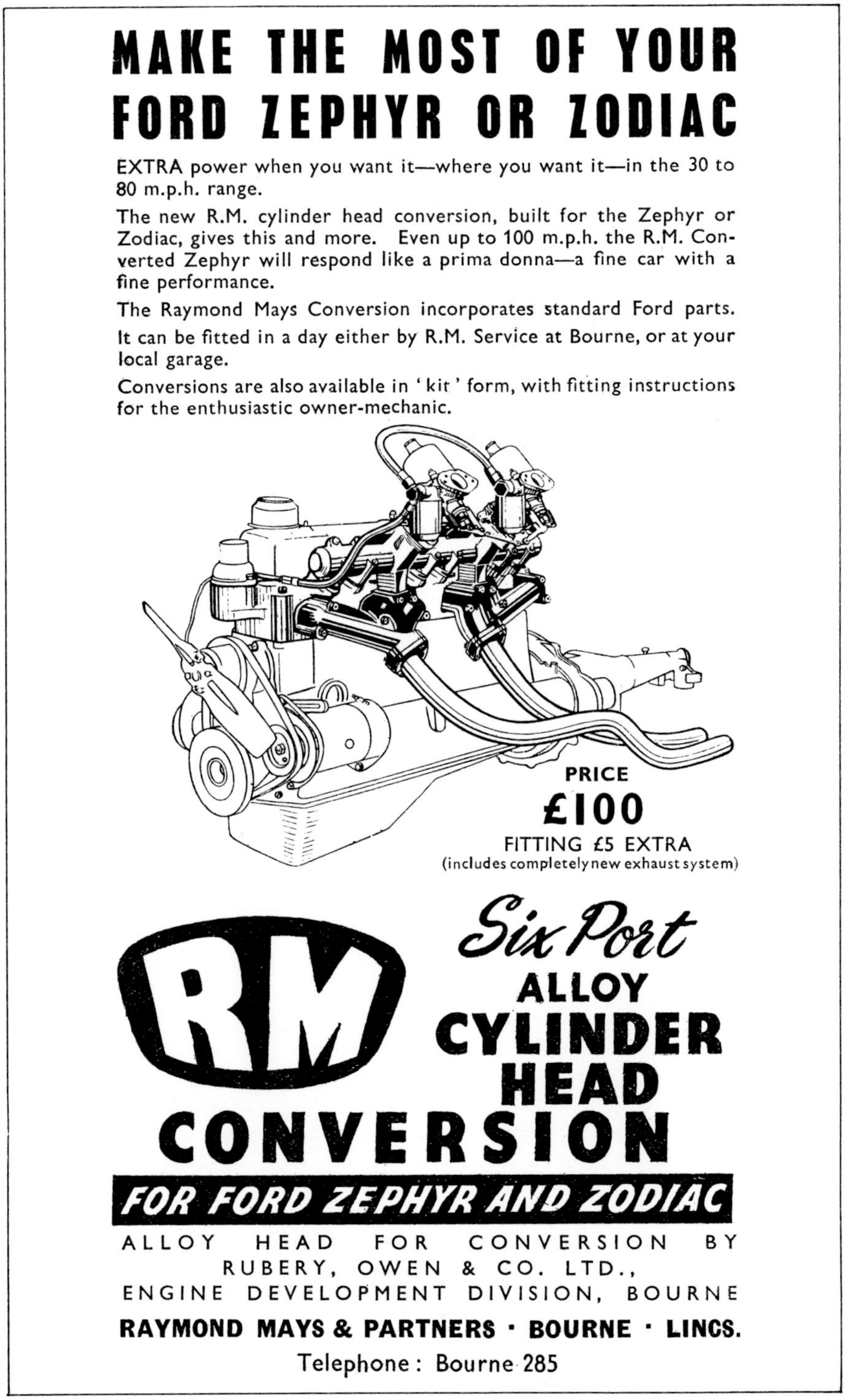

Recipe for 100mph; a Raymond Mays conversion. The earlier Mays conversion, based on the standard cast-iron cylinder head, featured triple side-draught SU carburettors, and to make room for these the battery had to be removed to a position in the boot.

An attractive way of increasing the luggage space was offered by this conversion, although situating the heavy spare wheel well outside the wheelbase would have an adverse effect on the handling qualities of the car.

Laycock de Normanville overdrive was being offered by Laycock Concessionaires G.E. Neville and Son, of Mansfield, with a fitting kit for the Consul/Zephyr range. The overdrive gave a 22.2% reduction in engine revs when engaged, which meant on the Zephyr 20.7mph/1,000rpm as against 16.1mph/1,000rpm on direct top gear. The Mays-tuned engine would pull just about 5,000 revs on the overdrive ratio, and so the true 100mph was exceeded slightly. For regular 100mph motoring Raymond Mays advised the fitting of the Laycock overdrive, although the engine conversion alone was available at £105, including a dual exhaust system and a fitting charge. The overdrive cost £72 extra, and also being offered by Mays was a set of Ferodo brake linings in racing material at £6 19s 6d, and stiffer rear shock absorbers at £6 6s.

John Bolster tested the Raymond Mays car complete with all the options, and reported his findings in *Autosport* on December 3, 1954. After describing the engine modifications, he went on: 'The car which I tried had a Laycock de Normanville device installed; this could be engaged on either second or top gear by a small switch on the gearlever, and gave the effect of a close-ratio five-speed gearbox. On the road, the first impression one gains is that the smoothness of running has been improved. This must be due to better fuel distribution, thanks to the new porting and manifold. During the tests, I changed from first to second at 30mph, second to overdrive second at 50mph, and then into direct top at 75mph. About 90mph seemed a good pace at which to switch in the overdrive on top'. Checked against the stopwatch the Mays Zephyr recorded 0-30mph in 3.5 seconds, 0-40 in 6.6, 0-50 in 9.4, 0-60 in 13.6, 0-70 in 17.8, 0-80 in 24.4 and 0-90 in 32 seconds. The standing quarter-mile had taken 19 seconds, and the maximum speed was a two-way average of 102mph. A tremendous overtaking ability was indicated by the time taken to go from 30 to 50mph in direct second gear of 5 seconds flat, and from 50 to 70mph in overdrive second in 8 seconds. The total price of the conversion, with the overdrive, hard brake linings and stiffer rear shock absorbers, would add £191 to the price of a Zephyr or Zodiac, giving a total price of £945 and £1,042, respectively, for which, in 1955, the buyer would have a truly outstanding high-performance saloon.

The cars in the Consul/Zephyr range had entered their fifth year of production firmly established as the leading sellers in their particular categories, and none of the more recently introduced competition was proving capable of changing the situation. The Consul, at £666, was faced with such opposition as the Austin A50 Cambridge, introduced in late

1954 at £678 and offering a similar level of performance and accommodation, the Vauxhall Wyvern at £702, which dated back to 1951 but had never seriously threatened the Consul's sales, and the Morris Oxford Series 2, which, at £744, was rather highly priced in this sector. The Zephyr and Zodiac, at £754 and £851, were enjoying a near monopoly of the family car six-cylinder market. Vauxhall's Velox, and latterly the Cresta, closely matched the Zephyr and Zodiac specification and were priced accordingly too, at £759 and £844, and yet had failed almost completely to capture the public's imagination in the way the six-cylinder Fords had done, and consequentially their sales were on a much smaller scale. The Austin A90 Westminster, a slightly larger and heavier car with a 2.6-litre six-cylinder engine, had been introduced in late 1954 and, at a price of £791, looked to be a very serious contender for the Fords. In the event, despite proving to be an excellent all-round car, it also failed to make a significant impact on Zephyr/Zodiac sales. Somehow, Dagenham had achieved a blend of qualities that simply made the Zephyr/Zodiac concept seem exactly 'right' for the vast majority of buyers in this sector of the market.

As well as having a wide public appeal, the Zephyr Six, with its high performance and, by now, very well proven durability, was finding increasing favour amongst the provincial police forces. Its rapid acceleration and good handling endeared it to the patrol drivers and, in one force at least, the Zephyrs had become affectionately known as the 'Fast Blacks'. A visit to the rider's car park at a league speedway track would often reveal a number of the big Fords. The professional speedway rider covered a high mileage in the course of an English league season, and a hard-working, dependable car, able if necessary to carry the speedway machine on a rack across the back, was a vital part of the rider's equipment. To see the 1954 World Champion, New Zealander Ronnie Moore, at the wheel of a Consul convertible, and England Captain Arthur Forrest with an always immaculate Zodiac, served to underline the esteem in which the big Fords were held. When the Hollywood Motor Rodeo toured England in the summer of 1955, they chose a fleet of Zodiacs to supplement the American Ford Fairlanes which they used in the United States to give their impressive display. The Zodiacs were an excellent choice, and put up some startling performances within the confines of the speedway stadiums chosen for the British tour.

A 1955 Consul convertible posing for a period press photograph.

The Autocar took another look at the Consul in 1955, not having put one through the full road test procedure since the original press demonstrator of 1951. The report appeared in the issue of July 22 and, four years on, *The Autocar* was still very favourably impressed, with comments such as: 'The Consul has considerable appeal, reasons for its popularity being readily apparent' and: 'Performance is impressive, the ride is comfortable, and the handling is such that the enthusiastic driver may, when he feels so disposed, throw the car into corners. . . . It is almost indifferent to rough surfaces'. Of the gearbox there was a hint of criticism: 'To use the gearbox well it is necessary to double-declutch when changing down into first, a manoeuvre unknown to many modern drivers, and one with which women usually have difficulty'. Summing up the Consul, *The Autocar* concluded: 'There is much about the Consul that is impressive, notably detail points of design that may not make themselves apparent under casual inspection. For carrying capacity, economy, performance and modest price and running cost, the Consul merits praise'. This particular car had accelerated from rest to 30, 50 and 60mph in 6.9, 17.2 and 25.9 seconds, and recorded a maximum speed of 74mph and an overall fuel consumption of 29mpg.

The success was being repeated on a worldwide basis in the various export markets. Sales of the Zephyr had received a boost in Africa in 1955 when Vic Preston drove a standard Zephyr Six to outright victory in the tough East African Safari Rally. In Australia the Consul and Zephyr had been well received from the start, and proved to be able to stand up well to the rough road conditions that existed in many areas outside the major towns. This aspect of the Zephyr's performance was reported on enthusiastically by Ian Fraser writing in the August 1955 issue of one of Australia's leading car magazines, *Cars Today*. According to Fraser: 'The Zephyr was very roadworthy in all conditions. On dirt roads the handling was nothing short of excellent. Belted along a corrugated gravel road at 80mph the car held on very well, with no tail hop at all. Driven over rough cattle tracks, the front springs refused to bottom, although one of the rear semi-elliptics went to the limit in a huge water-filled hole, which would have probably wrecked a lesser car'. He went on to criticize the Zephyr's 41ft turning circle, and thought that the driving position lacked support for persons of larger than average size, but he concluded: 'The Ford Zephyr is, without doubt, the leading car in its class. It has a couple of minor faults, but its advantages certainly outweigh these a hundred to one'.

Soon after ordering a new Zephyr Zodiac from Ford Agents Brown Brothers Motors, of Vancouver, 32-year-old Canadian Michael Schreiner learned that it would be Dagenham's millionth postwar export vehicle. Mr and Mrs Schreiner were flown to England, as the guests of the Ford Motor Company, to take delivery of their Zodiac at Dagenham. On September 5 the Zodiac was driven off the production line by The Rt Hon A. R. W. Low, Minister of State at the Board of Trade, and formally handed over to Mr Schreiner. After a tour of Dagenham in a cavalcade of Ford cars, which included a visit to the Civic Centre where the Schreiners met the Mayor, the new Zodiac was loaded on to the *MV Evanger* of the Norwegian Westfal-Larsen shipping line. The choice of the *Evanger* was an appropriate one, as, like the Schreiner's Zodiac, it too was making its maiden voyage. In the 10 years since the war had ended no other British motor manufacturer had matched Dagenham's million exports, which represented well over half of their total postwar production to that time of 1,800,000 vehicles. Just under a quarter of all Britain's export vehicles since the war had been

An interesting comparison between the Zephyr Six and the Zephyr Zodiac. The Zephyr Six shown has the optional radio. A radio exclusive to the Zodiac, featuring twin speakers, was available at £47.17s.1d.

products of the Ford Motor Company; these figures underline the tremendous contribution that Dagenham was making to Britain's postwar recovery.

The benefits to be obtained by adding an overdrive to the gearbox, particularly on the six-cylinder models, had been well demonstrated by the cars fitted with the Laycock electrically-actuated unit. Another overdrive for the Consul/Zephyr/Zodiac range was now available from Vehicle Developments, of London. Known as the Handa overdrive, this was engaged by a vacuum-operated diaphragm mechanism using manifold depression. Then, in August 1955, the Ford Motor Company themselves announced an optional overdrive for the six-cylinder models only, priced at £63. This was the semi-automatic Borg-Warner unit which incorporated a freewheeling arrangement effective below about 30mph. Above this speed the overdrive could be engaged in any gear simply by releasing the accelerator pedal momentarily: if direct drive was then needed above the 30mph cut-in speed, during an overtaking manoeuvre for example, the overdrive was disengaged by 'kicking down' the accelerator pedal past an extra switch at the fully open position, after which direct drive would be retained unless or until the pedal was lifted or the speed dropped to below the 30mph cut-out speed. In the under-30mph freewheeling range, transmission was through the ordinary gearbox ratios with normal drive on an open throttle. The combined overdrive/freewheel unit could be simply locked out of use altogether by means of a facia-mounted lever, leaving a normal three-speed transmission. The Borg-Warner unit gave a reduction of 30% in engine rpm, raising the mph/1,000rpm in top gear from the normal 16.1 to 23.0. As the six-cylinder engine was able to pull around 4,000 revs in overdrive, the Zephyrs and Zodiacs so equipped were genuine 90mph cars. Another, incidental, benefit of the Borg-Warner overdrive was that in the freewheel range, provided that the throttle was closed, clutchless gear-changes could be made, although of course the clutch still had to be used in the normal manner for stopping and restarting.

Also during 1955, the oil pump had received a modified pick-up pipe, making the pump self-priming, and the ignition system now featured an oil-filled coil. At the Earls Court Show that year there were just two visible changes: the Zephyr Six had acquired the Zodiac's vinyl roof lining, and all three models featured separate amber rear flashing indicators in place of those originally incorporated in the rear light units. Mounted in the back panel beneath the boot lid, the new indicators were a standard Lucas design and as such, unfortunately, did not quite blend into the rear-end styling of the car. The only other change in the specification was the adoption of tubeless tyres.

The 'Five Star Cars' were now nearing the end of their production run as, although their sales were steadily rising — a fact reflected in the record production figure of 106,176 cars for 1955 — development work and testing of the prototypes of the projected replacement models, the Mk 2 Consul, Zephyr and Zodiac, was now well advanced. When production of the Mk 1 cars ceased on February 22, 1956, which was the day after the Mk 2 had been announced, a total of 406,792 had been produced, this figure being made up of 231, 481 Consuls, 152,677 Zephyr Sixes and 22,634 Zephyr Zodiacs.

Surprise was expressed in many quarters that the Ford Motor Company had deleted the Mk 1 models whilst still in their prime, indeed, almost certainly before they had actually reached their peak annual sales potential. But, whereas in the past Dagenham had seemed quite content to offer the best value for money in any given class, even if the cars in question were somewhat dated, they were now keen to be offering the best cars as well, and with the Consul/Zephyr range they could justifiably claim to have done just that for the last five years. To maintain this position, particularly for another five years or more, it was essential to have not just a 'facelifted' car, but a genuine development, which, whilst still following closely the firmly established and very popular basic concept, would show real advances wherever possible in the light of experience gained with the original design, so as to put the names of Consul and Zephyr once more as far ahead of the opposition as the original models had been on their spectacular Earls Court debut in 1950.

CHAPTER 3

The Three Graces

Mk 2 Consul, Zephyr and Zodiac

With the introduction of the Zephyr Zodiac in October 1953, and production of the convertible models at last under way, the development of the Mk 1 range was more or less complete. The theme was firmly established, and sales of the entire range could obviously continue to expand for some considerable time with the absolute minimum of revision to either the appearance or the technical specification. Few, if any, totally new designs had entered production so smoothly, and with so few teething troubles as had the Consul and Zephyr Six; those of the original production batch still on extended 'life' tests had by now confirmed the remarkable durability that had been indicated in the tests carried out with the prototype cars, and so the Product Planning executives and the development engineers could now start to scheme out the Mk 2 versions of the Consul, Zephyr and Zodiac.

A larger car, one which would be a generous six-seater, was decided upon. Longer by 7in in both Consul and Zephyr form, and wider by 5in, the new bodyshell shared no panels whatsoever with its predecessors, although, apart from having welded-up rear wings, it still followed the same basic concept and constructional methods of the original model, and thus inherited the great strength and rigidity that the Mk 1 cars possessed. In fact, beam stiffness and torsional rigidity were greater by about 12% as a result of carefully redesigning the inner sill area to compensate for the increased length. Deeper windows and much slimmer windscreen pillars, combined with a fully wrap-round rear screen, gave a greater field of vision than on any previous British saloon car. The hooded headlamps and mildly finned rear wings could be seen from the driving seat by anyone of average stature, and these features, coupled with a much improved steering lock giving a turning circle of around 35ft, would enable these big cars to be handled with confidence in confined spaces such as car parks and driveways.

The layout of the luggage boot, which was now of 20cu ft capacity, remained the same as before, again including a toolbox in the Zodiac. The underslung fuel tank was enlarged to $10\frac{1}{2}$ gallons capacity and had a filler cap neatly concealed from view by the rear number-plate, which was mounted on a hinged flap. The large bonnet top was now counterbalanced, and when opened revealed a truly capacious engine bay. The underbonnet layout was altered slightly, with the battery changing

A 1956 Mk 2 Zephyr.

Mk 2 Consul. The chrome fillet on the rear screen pillar was introduced in October 1956, and has presumably been fitted retrospectively to this Consul, which has a February 1956 registration number.

sides and now occupying a position inside the offside front wing where the heater had previously resided; that unit was now placed on the scuttle bulkhead just above the height of the engine. The hitherto awkwardly placed fuel filter bowl was now detached from the pump body and occupied a convenient position in the fuel line on the offside inner wing. The Zodiac carburettor had a large oil-bath air filter as a standard fitting, whilst the two cheaper cars retained the washable-element type. The interior arrangements were planned around 'Oscar', Dagenham's 5ft 10in, 12 stone 'average man', whose measurements and weight were based on a detailed survey of 68,000 armed forces personnel, some 80% of whom had the average measurements which resulted in Oscar. Oscar existed in several forms, ranging from a whole series of transparent templates in various positions and scale sizes, up to a full-size and weight dummy figure.

The Mk 2 model exhibited a sharper profile than the superseded Mk 1 range.

Although the boot was of very ample proportions, the high back panel made loading inconvenient by comparison with rival makes, a point which had been criticized on the Mk 1 models. The fact that this criticism was not attended to on the Mk 2, by deleting the back panel, is an excellent illustration of Dagenham's refusal to sacrifice overall strength for the sake of a small convenience.

As before, the Consul and Zephyr furnishings were simple in style, and the Zodiac, now without the 'Zephyr' prefix, was trimmed and equipped in a more luxurious manner. All three cars featured a combined ignition/starter switch, an arrangement making its first appearance on a British car. Apart from this, the standard equipment of the Consul and Zephyr was very much the same as previously, but the Zodiac displayed some slight changes. A more elaborate tail-end styling treatment was employed which added a further 2in to its overall length, and at the front it sported a different grille with which to distinguish it from the cheaper Zephyr when viewed head-on. The auxiliary lamps and wing mirrors were deleted from the specification, but the two-tone paint, whitewall tyres, chrome wheelrim embellishers and gold-plated badges still made it instantly recognizable as the top-of-the-range model.

Offering so much more accommodation had inevitably resulted in an

Steering wheel, instruments and controls of the Mk 2 Zephyr/ Zodiac.

1956-57 Mk 2 Zodiac interior. The dished steering wheel was making its first appearance on a British car.

The optional nylonweave upholstery in an early Mk 2 Consul. The front seat centre armrest was not adopted for the production models at this stage.

increase in weight, although experience with the Mk 1 models had shown that they were in fact carrying excess weight and that savings could have been made without loss of strength. These lessons were applied to the new cars to keep the increase to a modest $\frac{3}{4}$cwt, which was also, of course, accompanied by an increase in frontal area. In order to overcome these two factors the engineers decided to adopt greater cylinder capacities for the Mk 2 engines, rather than resort to a higher state of tune or multiple carburettors. There was plenty of 'meat' in the block, and an increase in the bore to 82.5mm still allowed ample water jacketing around each cylinder to ensure even cooling. A new, slightly longer-throw crankshaft gave an increase in stroke to 79.5mm, so the over-square characteristics were retained with only a very slight increase in piston speed, which was in any case to be more than compensated for by the adoption of higher overall gearing. As before, the crankshaft ran in either three or four main bearings

This illustration shows the fresh air ducts on the bulkhead, which were fitted when the optional heater was not specified. The four-cylinder engine fits very comfortably in the Mk 2 Consul's engine bay. This engine and gearbox was also used in the Thames 10/12cwt forward-control van introduced in 1957. Although not a Ford option, the Perkins 1.6-litre diesel could be installed in the Consul, but as the Consul engine was unusually long-lasting by four-cylinder standards, very few owners bothered with the diesel conversion.

according to the four- or six-cylinder application, and to compensate for the longer throw the bearing journal diameters were increased to 2.376in and 2.125in, respectively, for the main and big ends. Hollow webs between the big ends made this an unusually light crankshaft, and in service it proved to be just as rigid as the shorter throw Mk 1 fitting. Separate valve guides were omitted from the new cylinder head, resulting in improved valve cooling coupled with a worthwhile cost saving. The wedge shape was retained for the combustion chambers, whilst the standard compression ratio was increased to 7.8:1. A 6.9:1 head for use with 'regular' grade fuel was to be available as an option. A similar capacity cooling system to the Mk 1 was retained, but now pressurized to 7psi. The new bore and stroke dimensions gave total cylinder capacities of 1,703cc and 2,553cc for the new Consul and Zephyr/Zodiac, and the power and torque outputs showed a considerable improvement over the Mk 1 cars; net figures were 59bhp at 4,400rpm with 91lb/ft torque at 2,300rpm for the Consul and 85bhp at 4,400rpm with 113lb/ft at 2,000rpm for the Zephyr and Zodiac.

A new clutch, with an $8\frac{1}{2}$in diameter lining, was introduced for the six-cylinder models, whilst the four-cylinder cars would continue to use a clutch of the same 8in diameter as before. The gearbox ratios were to remain the same, with first gear continuing to be unsynchronized. The gearbox had proved to be particularly trouble-free in the Mk 1 range, and so the only change in specification deemed necessary here was a slight increase in the diameter of the mainshaft in order to cope with the increased power from the enlarged engines. As the larger capacity engines would be capable of pulling higher gearing, all three gears were effectively raised by the choice of new rear axle ratios of 4.11:1 and 3.90:1 for the four- and six-cylinder cars, respectively, these ratios having been carefully calculated so that once again the maximum top gear speeds would coincide approximately with a piston speed of 2,500ft/min, the road speeds now being 79.5mph for the Consul and 88mph for the Zephyr and Zodiac. The new axle ratios gave a top gear mph/1,000rpm figure of 16.9 for the Consul (15.1, Mk 1) and 18.5 for the Zephyr/Zodiac (16.1, Mk 1) and so promised rather more effortless top gear cruising in the Mk 2 models with little, if any, increase in fuel consumption. With the addition of the optional Borg-Warner overdrive, the Mk 2 Zephyr/Zodiac gearing would be remarkably long-legged, with 26.4mph/1,000rpm offering better than 25mpg economy at a 70mph cruising speed. This device was considered at one stage as an option for the new Consul, in conjunction with the lower 4.444:1 axle ratio from the Mk 1 Zephyr in order to reduce the overdrive gearing to more manageable proportions for the four-cylinder engine. However, although listed in the original Mk 2 Consul brochures, it was apparently withdrawn as a Consul option soon after the model's introduction.

The suspension and steering set-up which had been pioneered so successfully in the Mk 1 range was, of course, retained, the only alterations being those necessary to accommodate the larger and heavier body. Brake drum diameter remained the same 9in as before, but the front shoe width was increased from $1\frac{3}{4}$ to $2\frac{1}{2}$in, giving a total brake lining area of 147sq in, a larger increase than was demanded simply by the weight of the new models in order to ensure adequate stopping power from the higher speeds which could now be attained. New road wheels, of the same 13in diameter and 4 or $4\frac{1}{2}$in rim width as before and retaining the 5.90 and 6.40 section tyres, were slightly less dished than those of the Mk 1 cars. The longer and wider body had naturally allowed an increase in the track and wheelbase, the

Mk 2 Zodiac underbonnet layout. The Zephyr retained a cylindrical, washable air cleaner rather than the Zodiac's oil-bath type pictured here. From October 1960, both six-cylinder Mk 2s featured a pan-type air filter with replaceable paper element.

latter in turn allowing the engine to be mounted a little further back in relation to the front axle line, so lessening slightly the nose-heavy bias in the front/rear weight distribution. The difference was most marked on the Consul, with a front/rear percentage figure of 55/45 compared with the 59/41 of both the four- and six-cylinder Mk 1 cars, the figure for the Zephyr now being 57/43.

The new models were given the factory designation 204E (four-cylinder) and 206E (six-cylinder) in right-hand-drive form, with 205E and 207E for left-hand-drive cars. The first prototype to be completed, a Zodiac, took to the road under the cover of darkness in December 1954, and by early January had been joined by another Zodiac and a Consul. Soon after running-in was completed, the three cars, accompanied by a standard Mk 1 Zodiac as a 'yardstick' for comparative purposes, left Dagenham for the North Sea ferry and then motored across Holland and into Germany, where a test headquarters had been set up in a remote mountain area. Within a few days a prototype Zephyr had joined the first three cars in Germany to take part in what was later described by Ford as 'one of the most intensive proving programmes in automobile history'. Germany offered almost ideal facilities, with many miles of badly made-up roads and shattering *pavé,* in addition to the autobahns where full-throttle driving could be indulged in for miles at a stretch, whilst overnight temperatures in the mountains of around −40°F were also very useful for evaluating cold start performance. Meanwhile, other prototypes were undergoing tests at Ford's own proving ground in Britain, whilst additional testing, principally of the suspension, was being carried out on the uniquely rough roads of Kenya, despite the problems being caused there at that time by the Mau Mau terrorists.

By late 1955 much of the development work was complete, and a production date was set for the following January. Pilot production of the Zephyrs and Zodiacs got under way on January 26, with the first Consuls following early in February. On Tuesday, February 21, 1956, 'The Three

Graces' — Dagenham's Mk 2 Consul, Zephyr and Zodiac — were officially announced. The Ford Motor Company hired Harringay Arena for the following weekend, to stage a lavish showing of the new cars for the benefit of Ford employees, dealers, overseas Ford officials and VIPs. In the arena, 'The Three Graces' were set in front of a background of apple blossom and cherry trees, and were surrounded by exotic flowers flown in from Egypt and the Continent.

The styling of the new cars was greeted with tremendous enthusiasm by both press and public alike, and in the opinion of many people, Dagenham's styling team, led by Colin Neale, had produced the most handsome cars ever to appear from a British manufacturer. A strong transatlantic influence was evident once again, the new cars displaying a subtle blend of features from the parent company's 1954/55 Mainline/Customline/Fairlane series, the extra length with no increase in height giving a lithe or 'graceful yet athletic' appearance as stylist Colin Neale put it. Complementary to the new appearance was a range of colours specially blended for 'The Three Graces' consisting of Carlisle Blue, Sarum Blue, Corfe Grey, Wells Fawn, Hereford Green, Warwick Green, Ivory and Black, with five two-tone combinations of these, plus Black, being available on the Zodiac.

Despite the cost of retooling and the increase in the raw materials necessary for the larger body, the new models were once again offered at prices consistent with the Ford tradition of unbeatable value for money. With the Consul listed at £781, the Zephyr at £872 and the Zodiac at £969, it was obvious from the outset that the Mk 2 models would continue to build on the remarkable sales success of their predecessors and would further enhance the excellent reputation which now came with the names Consul, Zephyr and Zodiac. Closely following the saloons into production were convertible versions of the Consul and Zephyr, the former priced at £946, with a power-operated hood as an option for an extra £75. The Zephyr convertibles were built with the power tops as standard and were priced at £1,111, although a manual top version was available to special order with the corresponding £75 reduction.

A road test report of an overdrive-equipped Zephyr appeared in *The Autocar* on April 13. In every respect the car came in for considerable praise, with comments such as: 'First impressions of the engine on the road were most favourable . . . The spirited fashion in which the car tackles main road gradients and overtakes slower moving vehicles is most exhilarating for the driver . . . The brakes are of Girling hydraulic type; they could not be faulted by heavy use and extreme pedal pressures. There was no unevenness or overheating, and only slight fade after exceptionally heavy use during track testing . . . The rear end is no longer skittish, nor does it attempt to break away in an over-fast corner. In the wet and with increased power, there is naturally more likelihood of the tail coming round, but not, we are satisfied, more than with any similar car'. Acceleration through the gears showed a worthwhile improvement over the Mk 1 Zephyr, with 0-30, 0-50, 0-60 and 0-70mph being recorded in 4.9, 12.3, 17.9 and 24.5 seconds, although the maximum speed of 86mph suggested that this particular Zephyr was not yet fully run-in. With the overdrive in use 32mpg was recorded on one run when the car was being driven with economy in mind, and the overall consumption figure quoted was a creditable 24.1mpg. One of *The Autocar* staff had confessed to being 'well-nigh electrified by the excellence of the new Zephyr', and the test was summed up with the comment, 'it is one of the best and most encouraging British cars in large-scale production that we have tested since the war. And there is no reason to qualify that statement'. In the hands of *The Motor*, (R/T No 9/56) a new Zodiac recorded maximum speeds in the three gears of 36, 62, and 90.5mph, and accelerated from rest to 30, 50, 60 and 70mph in 4.6, 11.3, 17.1 and 24.3 seconds, and as this particular car had already run upwards of 2,000 miles these figures were more representative of what could be expected from a fully run-in example.

A worthwhile improvement in all-round performance was also evident in the new Consul. An example reported on in the June 15, 1956 issue of *The Autocar* reached 79mph and accelerated from rest to 30, 50 and 60mph in 6.4, 15.8 and 25.0 seconds. With second gear now comfortable up to 50mph or so, the Mk 2 had a big advantage over the early Consul when overtaking in the 30-50mph bracket, with a very useful time of 8.7 seconds being recorded for this range in this gear. An overall fuel consumption of 29mpg over a distance of 540 miles suggested that the performance increase was not at the expense of economy. *The Autocar* reported: 'This new Consul is attractive in its class as it offers so much room and performance, at a price of well under £800. It combines a handsome modern appearance with technical improvements of real value, and is likely to remain for a long time in the forefront of this country's best buys'.

'The Three Graces' made their North American debut at the 1956 New York International Motor Show, and only eight minutes after the doors opened on the first day, April 28, two Ford salesmen were each claiming to have made the first sale of the show. Show Manager Gerald Martin declared a dead heat, announcing that Hans Hallendorn had made the first sale of a car for delivery in the United States, a Consul, whilst the order for a Zephyr taken by Horace Sirraco was the first from an American who wished to take delivery in Europe under a plan enabling the buyer to use his new car on a European holiday, then have it shipped to America by Ford at no extra cost. More than 50% of the Mk 2s being produced were going for export, but this flow suffered a temporary setback in July because of an industrial dispute within the rival BMC organization. Fisher and Ludlow, of Birmingham, who were a part of BMC, but also provided parts for other manufacturers, were at that time supplying Briggs Motor Bodies with certain panels for the Mk 2 range, and a strike there halted these supplies. The dies used by Fisher and Ludlow were Ford property, but attempts to retrieve them were baulked by Transport Union drivers in Birmingham refusing to handle them, even though their own Union officials tried to persuade them otherwise. To combat this situation and keep the men at Dagenham employed, Ford hastily recalled as many as possible of the 'knocked down' export cars, including those already packed and awaiting shipment at the docks.

Some of the early production six-cylinder Mk 2s had suffered from broken halfshafts, and a modification to cure this was introduced in September. A new 24-spline shaft, with the appropriate differential gears, replaced the original 16-spline shaft, an arrangement which had been inherited from the Mk 1 model and apparently could not always withstand the increased torque from the larger Mk 2 engines. The modification applied to both four- and six-cylinder cars and completely eliminated the problem.

The new models made their first Earls Court appearance in October 1956, on stand 145. Two very slight styling changes were evident; a chrome-plated fillet had been introduced on the rear screen pillars, and the flat panel across the rear of the car beneath the boot lid now featured two moulded ribs. A Zodiac convertible was being shown for the first time, and announcing this glamorous addition to the range the Ford Motor Company said: 'The distinction of the Ford range attains a new peak with the introduction of the Zodiac convertible, on which a power-operated hood is standard equipment. In this beautifully styled car is incorporated all the splendour and refinement of the Zodiac saloon plus, when required, the benefits of open car motoring. This truly superb example of convertible design has a basic price of £835 and with purchase tax costs a total of £1,253 17s'. Also being shown on the Ford stand was a Consul 'Farnham' estate car, and on their own stand, 111, the coachbuilders E. D. Abbott, of Farnham, were showing estate car versions of all three Mk 2 models. The conversion closely followed in style that of the successful Mk 1 estates and gave a very useful 36cu ft of luggage space, or 66cu ft with the rear seat squab folded flat. Stiffer rear springs and oversize tyres, 6.40 × 13 on the Consul and 6.70 × 13 on the Zephyr/Zodiac, were now included in the estate car specification. Prices were £1,028, £1,118, and £1,223 for the Consul, Zephyr and Zodiac. A new trim material, nylonweave, was announced as an optional extra on the Consul and Zephyr to replace the standard PVC, and as a no-cost optional alternative to the standard leather upholstery in the Zodiac.

Much more significant, however, was the announcement that a Borg-Warner automatic gearbox was available on the six-cylinder saloons and estates, and at a price of £187 which would make a Zephyr saloon so equipped the cheapest automatic car available outside the United States. The gearbox chosen was the Model DG; a robust unit designed to handle the torque from engines of over $3\frac{1}{2}$-litres capacity, it would be under no stress whatsoever behind the $2\frac{1}{2}$-litre Ford engine, thus promising a long trouble-free life. An automatic Zodiac was soon made available to the motoring press, and a report of the car appeared in the February 8 issue of

The Mk 2 'Farnham' estate car. The rear screen is still the Mk 1 item.

Dagenham's and, some insist, Britain's most handsome convertible, the Mk 2 Zodiac.

The Autocar. The car accelerated from rest to 70mph in 31.6 seconds, some 6 seconds longer than the time they had recorded with a manual Zephyr. This performance decrease could be accounted for by the power losses inherent with a torque converter in place of a clutch, and by the increase in weight — about 1½cwt — of the automatic box. The fuel consumption, with an overall figure of 23.1mpg, was little different from that of a manual Zephyr, and was helped no doubt by the fact that on the DG gearbox the torque converter only operated on low and intermediate ratios and was locked out in top gear. In addition the automatic Zephyrs and Zodiacs were fitted with larger 6.70 section tyres in view of the extra unladen weight, and were consequently pulling slightly higher gearing. *The Autocar* summed up: 'As our experience of this latest line from Dagenham broadens, the value for money, combined with roadworthiness of a high order, becomes even more impressive. This automatic version of the Zodiac II demonstrates that a 2½-litre engine so equipped can have a very acceptable performance and restfulness, for those who consider the normal shift to be outdated'.

The introduction of the Mk 2 in Australia quickly led to a revival of the plans for a pick-up version. The design team at Geelong produced full-size designs in 1956, based very closely on those of the ill-fated Mk 1 pick-up

Her Majesty the Queen's special-bodied Zephyr estate car.

Above left, automatic transmission for Her Majesty. Above, rear passenger compartment of the Royal estate car. Enlarged window frames are of timber construction.

Tailgate detail of the Royal estate car.

One of the extremely rare Dagenham-built pick-up trucks.

The Australian-built Mk 2 pick-up truck featured a different rear cab panel from the Dagenham version; the rear window is from the Mk 1 saloon. This Consul was photographed in 1983. Photos by Dick Barry.

project. One of the team, Lewis Bandt, took the drawings to the parent company in the United States early in 1957 and received authorization to tool-up for production. The prototype was tested with a 9cwt load over a 4,000-mile route in company with a Ford Mainline V8 pick-up before production commenced. Meanwhile, rather curiously in view of their original objections, Dagenham had built a Zephyr pick-up truck in November 1956, differing only slightly in detail from the Australian design, and production records show a total of 46 of these being built at Dagenham between then and August 1959; apparently these were all shipped to Australia.

Despite the Suez crisis, and the subsequent period of petrol rationing, worldwide sales of the Mk 2 range expanded rapidly during 1957, and by the middle of the year production was running ahead of the peak rate achieved by the Mk 1 models in 1955. Nevertheless, Dagenham planned several improvements designed to keep 'The Three Graces' out in front, and at Earls Court in October a Consul De Luxe was unveiled. Priced at £871, the De Luxe was a particularly useful addition to the range as, apart from the fact that the heater was still only an option, it was otherwise trimmed and equipped almost to the same standard as the Zodiac, with such items as a cigarette lighter, twin horns and windscreen washers, in addition to the leather or nylonweave upholstery, and so appealed to those buyers who appreciated the luxury trim and fittings, but had no desire for the high-performance six-cylinder engine. Externally the De Luxe model was identified by a contrasting colour applied to the roof, and chrome-plated surrounds to the rear lamps. As slight visual distinction on the Zephyr was apparent in the form of a new grille and the adoption of the Zodiac sidelights at the front, whilst the Zodiac itself remained unchanged externally. A redesigned front seat was a feature of all three cars, and both front and rear seats now also had the benefit of foam rubber padding in addition to the springing. A front seat centre armrest was also at last making its appearance, although still only as an optional extra on the Consul, and on the Zephyr and Zodiac at the expense of the one previously

Not quite right. A proposed grille design for the 1958-model Zephyr.

in the rear. Technical changes were the introduction of a Burman recirculating-ball steering box, and a new gear-change mechanism enclosed in the steering column. Completing the updated specification was a totally new range of 15 colours, including a red and a yellow, replacing the eight original shades of 1956.

For those owners wishing for a faster than standard car, the tuning specialists were offering similar conversions to those which had proved so successful on the Mk 1 models. The Alexander Engineering Company's conversion for the Zephyr consisted of a reworked cylinder head of 8.7:1 compression ratio in which the valve ports were opened out, larger inlet valves installed, double valve springs fitted and the combustion chambers polished. A new inlet manifold with triple SU carburettors supplied the mixture, and a six-branch exhaust manifold completed the conversion. The figures claimed by Alexander make an interesting comparison with those obtained by *The Autocar* for a standard Zephyr Mk 2.

Acceleration from rest (seconds):	Standard	Alexander
0-30mph	4.9	3.1
0-50mph	12.3	8.0
0-60mph	17.9	10.7
0-70mph	25.4	14.8
0-80mph	35.8	22.5
0-90mph	—	29.2

Acceleration in top gear (seconds):	Standard	Alexander
20-40mph	8.0	6.7
30-50mph	8.8	7.2
40-60mph	10.0	8.0
50-70mph	12.6	8.5
60-80mph	17.4	11.8

Inevitably, Raymond Mays developed a Mk 2 six-port light-alloy cylinder head, incorporating larger valves with double valve springs and giving a compression ratio of 8.7:1. Complete with twin SU carburettors and a dual exhaust system, this conversion, costing £135, put the Zephyr into the 100mph-plus category and enabled it to accelerate from rest to 30, 50, 60, 70, 80 and 90mph in 3.3, 7.5, 10.0, 14.5, 19.3 and 25.5 seconds, respectively, when tested by *The Autocar*. Later, Mays was able to improve the performance still further with a 9.2:1 compression ratio, triple SUs and a high-lift camshaft.

Servais Silencers produced a complete exhaust system for the Mk 2 models, including the convertible, which, because of its additional underbody bracing, required a different system from that of the saloons. Tests carried out by Servais with a Zodiac convertible resulted in the following 'before' and 'after' figures.

Acceleration from rest (seconds):	Standard	Servais
0-60mph	16.1	14.4
0-70mph	21.3	20.0
0-80mph	30.5	26.5

Acceleration in top gear (seconds):	Standard	Servais
40-60mph	9.7	8.2
50-70mph	11.2	9.9
60-80mph	20.2	16.3

For the vast majority of owners, the performance of the standard models in both four- and six-cylinder form was, of course, entirely satisfactory, and converted cars only represented a very tiny percentage of the total. There was, however, a popular misconception at that time that, because the Consul was lacking a 'third' gear, its performance was handicapped. In fact, the Consul actually had a performance advantage over some similarly-priced rivals of similar power/weight ratio, but equipped with four-speed gearboxes. The Hillman Minx, for instance, had a third gear ratio which gave 10.2mph/1,000rpm, and accelerated in this ratio from 30 to 50mph in 10.6 seconds (*The Autocar*, July 1956). The Morris Oxford Series III had a third gear giving 10.0mph/1,000rpm, with which it covered the 30 to 50mph range in 9.9 seconds (*The Autocar*, October 1957). The

Almost there. The grille design chosen for 1958 on, but seen here with a badge which did not see production.

1958 Zephyr. The grille on this car displays the badge which was chosen for production.

Consul was almost exactly similarly geared in its second gear, 10.2mph/1,000rpm in this ratio enabling it to cover the 30 to 50mph range in 8.7 seconds (*The Autocar*, June 1956). Indeed, if we look at the second gear ratios of these two popular rivals, we find gearing of 6.2mph/1,000rpm in both cases, which is only very slightly higher than the Consul's 5.95mph/1,000rpm bottom gear. So these two four-speed cars had, in effect, a three-

Prototype Consul De Luxe, 1957-58 version.

speed gearbox with an additional extremely low, and largely superfluous, bottom gear. There were, of course, some excellent four-speed gearboxes around at that time, but there were also far too many of the type just illustrated, of which the extra cost in comparison with the Ford Consul/ Zephyr box was a complete waste of money. Being the smaller-engined car of the range and sharing the same six-seater body with the six-cylinder Zephyr, would sometimes lead to the Consul being saddled with an 'underpowered' tag. Yet again, a look at contemporary road test reports is all that is necessary to refute this argument. The Wolseley 1500, introduced by BMC in 1957, had a 50bhp version of their 1,489cc B series engine to cope with a weight of 18cwt, and was fitted with a four-speed gearbox generally considered to have excellently chosen indirect ratios, allowing over 45mph in second and just over 70mph in third gear. During 1957, *The Autocar* tested both the new Wolseley and the Consul. The Wolseley was tested at a laden weight which gave it a power/weight ratio of 47bhp/ton, whilst the Consul had the remarkably similar figure of 46.8bhp/ton as tested. Of the Wolseley, *The Autocar* said: 'It is eminently suitable for the keen motorist who must compromise between sporting and business use'. It had accelerated from rest to 50, 60 and 70mph in 16.1, 24.4 and 43.9 seconds, and had recorded a mean maximum speed of 77.5mph with a best run of 79.2mph. The Consul, with fractionally less power/weight, and a three-speed gearbox, had gone from rest to 50, 60 and 70mph in 15.6, 25.4 and 38.9 seconds, and also recorded a mean maximum speed of 77.5mph, but with a slightly improved best speed of 80.6mph. Over the standing-start quarter-mile the Wolseley just beat the Consul with a time of 22.1 seconds against the latter's 22.7. These results show that the Consul was in fact extremely competitive, and also that within an 80mph performance range it was possible to tailor a three-speed gearbox perfectly satisfactorily to an engine of modest capacity.

Several interesting Mk 2 models came under the close scrutiny of the test staff of the various motoring magazines during 1957/58. In their issue of August 2, 1957 *The Autocar* featured a Zephyr estate car, and commented very favourably on the versatility of the Abbott conversion. Points earning praise were that the rear seat squab could be quickly detached and conveniently secured to the back of the front seat, and that the rear seat cushion and squab 'can readily be removed for use outside the car, for example, when picnicking'. When heavily laden, the Zephyr estate car had continued to handle well, *The Autocar* saying that 'we can reassure the prospective owner of a Zephyr estate car that the handling qualities are less affected by load than are those of most vehicles of this class'. A portion of the test mileage was carried out with a 6cwt burden on the rear platform in addition to two occupants, and *The Autocar* commented: 'The effect on acceleration was hard to detect; braking power was sufficient to cope with the load, even when the overdrive permitted the freewheeling below about 30mph. Steep hills were climbed without the impression that more power would be useful'. The overdrive had enabled the Zephyr estate car to record a commendable 25mpg overall for a distance of 682 miles, with a best figure of 33mpg for an 80-mile run with just two people on board.

The Motor sampled a Consul convertible (R/T No 2/58) and felt that 'it does not conceal its relationship with Detroit-built automobiles of the same name', going on to say that 'in what it does and how it does it, this medium-powered British car with an American "air" proved itself during an extended test to be a very desirable property'. Although tested in the winter months, this convertible apparently gave *The Motor* staff some enjoyable open-air motoring, as is indicated by the comment: 'As an open car, this convertible model proved very pleasant on fine winter days in the country, being immune from the buffeting around the driver's head, which can make long fast drives in some open cars very tiring'. In its closed condition, however, this particular car was apparently not entirely without draughts, and there was also criticism of the continued absence of sun visors in the Ford convertibles.

John Bolster put a Zodiac automatic through its paces, his report appearing in the June 13, 1958 edition of *Autosport.* He was particularly

1957-58 Zodiac Automatic. This particular car was tested by John Bolster for the magazine Autosport.

pleased about the driving position, having previously driven an early Mk 2 in which the pedals and seat had apparently not been in the correct relationship to each other, and he now stated: 'On taking my seat, I was at once delighted to find that the driving position has been vastly improved. Indeed, I was almost rude about this feature on the previous occasion, and it gives me very great pleasure to be able to withdraw the criticism now'. About the performance of the automatic car he continued: 'Borg-Warner gearboxes are specially tailored to give the best performance on the engine to which they are to be attached. In the case of the Ford Zodiac, a particularly felicitous marriage has been achieved, and the car is much more pleasant to drive than most automatics. The performance figures are

1958 Zodiac Automatic, with optional radio.

1958 Consul De Luxe.

just a little below those of a similar car fitted with a manual box and handled by an expert. However, the average driver would probably achieve a better all round performance with the Borg-Warner transmission. There is a Low position on the gear-lever for engine braking under difficult conditions, and a Kick-down on the accelerator to make second speed available for rapid overtaking. The engine has plenty of "punch" at all speeds, and this is a lively car . . . There is something very attractive about the easy way in which the car covers the ground, and this is a most untiring vehicle for long journeys'. Of the roadholding and handling, he wrote: 'The machine can be cornered quite fast, and remains fully controllable on wet roads. The steering feels positive, and gives the driver a sense of mastery over his mount'. John Bolster's general comments included the observations: 'The Ford Motor Company have been making cars of this type for a considerable time. The experience has been applied with considerable advantage to the latest model, and it is a thoroughly well thought out design from the point of view of all occupants'. He concluded his report with the comment: 'The man in a hurry will like this car, and so will the lady of the house who uses it for shopping. Many a family will get very fond indeed of their Zodiac automatic'. Checked against the stopwatch, this car had recorded times from standstill to 30, 50, 60 and 70mph of 5.0, 11.8, 17.4 and 28.4 seconds, respectively. The standing-start quarter-mile had occupied exactly 21 seconds and the maximum speed was 87mph.

In Australia, sales of the Zephyr coupe utility, as the pick-up truck was known, were fully justifying the faith shown in the design by the Ford engineering team at Geelong, and a test report on the 'ute' appeared in the Australian motoring magazine *Wheels* in May 1958. The design included a

Facia panel and controls of the automatic Zodiac. The dip-switch is to the left of the brake pedal and, on production automatics, was fitted with an organ pedal similar to the accelerator. The wide brake pedal was to allow comfortable left-foot braking.

stiffened rear chassis/floor and the addition of three extra leaves in the rear springs. The front end, including the passenger compartment, was standard Zephyr. The 'ute' was tested with sand bags on the rear deck, bringing it up to 33¼cwt, and at this weight it accelerated from rest to 50mph in 16.3 seconds and continued to a maximum speed of 77.2mph. The fuel consumption with this load varied between 20.3mpg 'hard driving' and 26.4mpg 'normal fast cruising'. As was usually the case with the Zephyr saloon when on test, the handling was highly praised, with the comment that 'we quickly became more confident of the Zephyr's handling qualities than we are of some cars deliberately designed for the sporting motorist!'. It was felt by the test team, however, that the normal

Mk 2 Consul convertible. The wing mirrors were a standard fitting on this model. This particular Consul was road-tested by The Motor.

1958 Zephyr convertible with optional whitewall tyres.

saloon gearing employed in the 'ute' was rather too high for this type of vehicle, particularly for fully-laden hill starts.

As the 1958 Earls Court Motor Show drew closer, large newspaper advertisements for the big Mk 2 Fords were appearing, headed 'Unchanged — Unchallenged', signifying that the Consul, Zephyr and Zodiac models on the Ford stand this year would be just the same as last. With total Mk 2 production by now running at a rate of over 2,400 per week, and at that still not always quite managing to satisfy the demand, there was certainly no need for any cosmetic changes with which to boost sales appeal at this stage. In fact, though, behind the scenes at Dagenham work was in hand on a 'facelift' for early the following year, designed to put the range in an unassailable position for the immediate future, and so allow development to begin on an eventual replacement model.

All three Mk 2s were dominating their particular market sector. The Consul, at £818, or £871 in De Luxe guise, was faced with the BMC Austin A55 Cambridge at £808, or £856 (De Luxe), a slightly smaller car of rather less all-round performance. The other BMC offering in this category was the Morris Oxford Series III, which at £884 was certainly trimmed to De Luxe standards, but had a distinctly lower performance than the Consul and rather more dated styling, and so this two-pronged attack by BMC was not having a serious effect. Standard were offering the Ensign, a smaller-engined (1,670cc) and more utilitarian version of their well known Vanguard, but at a price of £899 it was not competitive at all in this sector. Vauxhall had decided simply not to risk facing the Consul head-on, and had replaced their large four-cylinder car, the Wyvern, with the much smaller Victor in an attempt to pick up sales lower down the market. The Zephyr was a winner in every respect; priced at £916 it was unbeatable value in the family car six-cylinder class. Its prestige regularly boosted by rally successes, it had continued to build on the Mk 1 Zephyr's established reputation for consistent performance and durability to such an extent that only one competitor, Vauxhall, were still offering an under-£1,000 six-cylinder car. This was the PA series Velox, which at £983 gave similar accommodation to the Zephyr and almost as much straight-line performance, but still somehow lacked the correct blend of qualities to lift it out of the ordinary. At £1,013, the Zodiac had to contend with the very similarly equipped Vauxhall Cresta at £1,073, and the basic Austin A95 Westminster at £1,043. A fully equipped Westminster, the A105, was also

Today, the most collectible Mk 2 of them all, the luxurious Zodiac convertible.

available, and its specification included twin carburettors for the 2.6-litre six-cylinder engine, but at £1,235 it was not a serious threat to the Zodiac, although a close look at either of the Westminsters would indicate that if BMC could have reduced their prices to nearer the Zephyr/Zodiac level without sacrificing quality, the Westminster may well have enjoyed a much wider popularity.

On February 25, 1959, three years and four days since 'The Three Graces' had made their debut, Dagenham announced the 'Low Line' Mk 2 models. A new panel, approximately $1\frac{1}{2}$in lower and considerably flatter, replaced the original, slightly domed roof pressing, giving a surprisingly sleeker appearance which was further enhanced and highlighted by additional bright metal embellishment. All three cars acquired stainless steel windscreen and rear window surround cappings which completely covered the window rubbers whilst a capping of similar material was also applied to the roof drip rail. New chrome-plated headlamp hoods replaced the original painted fittings, and all three cars received restyled chrome plated rear lamp covers. The Consul was given the chromed door window surrounds and pointed rear wing tips which had always been a feature of the two six-cylinder Mk 2s and, along with the Zephyr, it inherited the extended rear wing side strips of the Zodiac. The Zephyr's grille badge was now also applied to the Consul De Luxe, and the latter featured new wheel trims in anodised aluminium, which were also replacing the chromed rings on the Zodiac's wheels. Exclusive to the Consul De Luxe were chrome cappings to the ribs along the back panel.

The interior had undergone a complete redesign, and current thoughts on safety had led to the introduction of a foam-padded facia and instrument cowl and soft, crushable sun visors, these latter items at last also making their appearance in the convertible. The steering wheel had always been deeply dished on the Mk 2, and the only changes here were the adoption of a black finish and, where applicable, a half horn ring in place of the previous full circle. Door and seat trim patterns were new, and all

1958 Zephyr estate car. The two-tone paint division shown was applicable to estate cars only.

1958 Zodiac estate car. The off-side wing mirror was a standard estate car fitting. Zodiac estates were also available with a 'sandwich' two-tone scheme.

models now had door pull-cum-armrests front and rear, those in the front being adjustable on the Zodiac. A map pocket on the front interior trim panel under the dashboard was a new feature which was confined to the Consul De Luxe and Zodiac, whilst the control knobs on all models displayed internationally recognized symbols in place of the previous capital letters for identification. The standard PVC trim material of the Consul and Zephyr displayed a completely new grained texture, and describing this the Ford Motor Company said: 'This material represents a departure from the traditional practice of using PVC to imitate cow hide. The Company's stylists regard PVC as a first class upholstery fabric in its own right and have adopted the Cougar Grain finish which has an individual and most attractive texture'. A further optional trim material, rayonweave, was also now available.

Prototype Zodiac two-door pillarless fixed-head coupe. A design which was definitely more than just a styling exercise, but which, unfortunately, did not reach production. Dagenham was already working to capacity, and an additional variation of the range, with only a very low-volume sales potential, could not really be justified.

Below, the projected hardtop featured a very transatlantic looking interior. Below right, power for the two-door hardtop was apparently to have included a triple Zenith carburettor arrangement, similar to that used on certain Mk 2 Zephyrs for the Alpine Rally.

A development Consul, featuring the 'Low Line' roof panel, but retaining the earlier screen surrounds and the early Mk 2 Consul cut-back fins.

Technical changes included the readoption of the original exposed gear-change linkage, although now shrouded in a similar fashion to that of the Mk 1 cars. A twist release handbrake was new, as was a front door safety locking mechanism, which rendered the interior door handle inoperative if the locking button on the door window sill was depressed. A final useful improvement was positive-locking quarter-lights which could not be opened from outside the car.

In all, these changes represented a worthwhile improvement in many respects, allied to a clearly updated appearance and, as they were incorporated at no extra cost, Dagenham's grip on this sector of the market was even tighter than before. Sales continued to expand steadily both at home and abroad following the introduction of the 'Low Line' models and,

'Low Line' Zephyr. The flatter roof panel gave an appreciably sleeker appearance and was accompanied by very slightly deeper front and rear screens than on the earlier Mk 2 models.

The rear-three-quarter view of the 'Low Line' Zephyr.

The glamorous 'Low Line' Zodiac.

The 'Low Line' Consul De Luxe; Zodiac luxury allied to four-cylinder economy.

Three-quarter-rear view of the Consul 'Low-Line'.

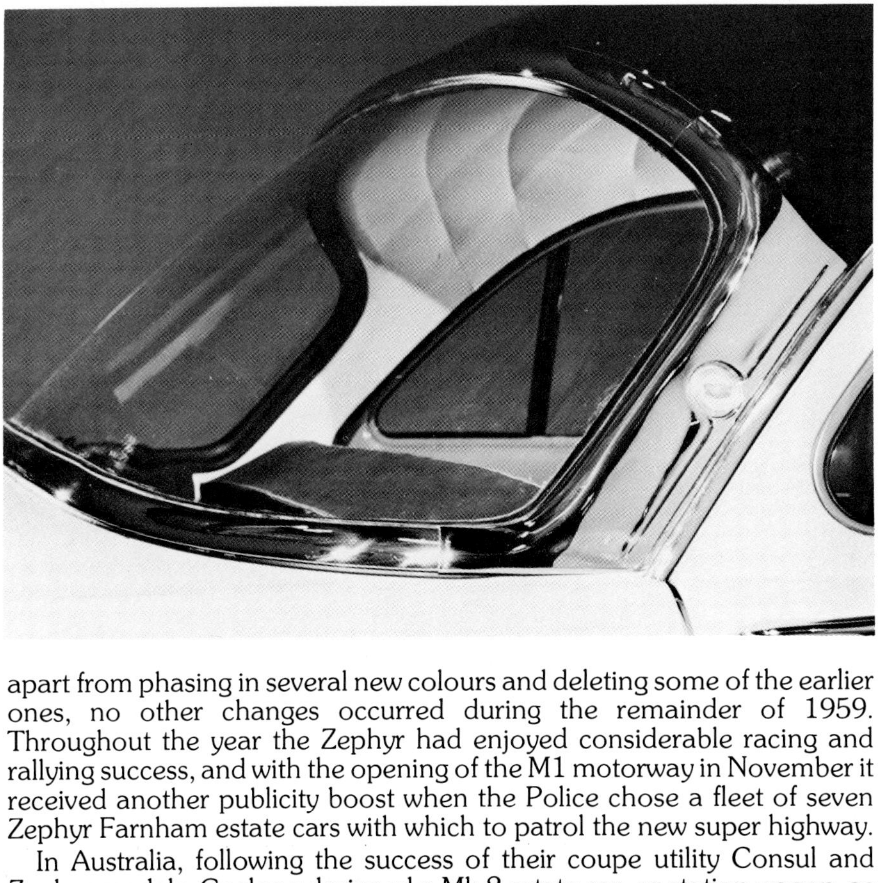

The front and rear screen surrounds of the 'Low Line' cars were in a rustless chromium alloy.

apart from phasing in several new colours and deleting some of the earlier ones, no other changes occurred during the remainder of 1959. Throughout the year the Zephyr had enjoyed considerable racing and rallying success, and with the opening of the M1 motorway in November it received another publicity boost when the Police chose a fleet of seven Zephyr Farnham estate cars with which to patrol the new super highway.

In Australia, following the success of their coupe utility Consul and Zephyr models, Geelong designed a Mk 2 estate car, or station wagon, as they preferred to call it, which, unlike the Abbott conversion in Britain, was rather more of a purpose-built job to cope specifically with Australian outback conditions when necessary. Although retaining the saloon rear wings, the rear end was otherwise totally different. The rear passenger

'Low Line' Consul head and sidelight detail. Shown is the production chromed headlamp hood, which was slightly less sharp than the earlier painted hoods. Some 'Low Line' prototypes featured headlamp hoods which were chromed versions of the earlier shape.

doors had squared-up window frames and the roofline at the rear continued horizontally, van-like, to the end. Whereas the Abbott estate rear door opened sideways, the Geelong wagon featured a very neat tailgate, hinged at the bottom, which was lowered after the pillarless rear window was first wound down into it by means of a lockable handle. Inside, the front bench could be moved forward up to the steering wheel, and when in this position, with the rear seat folded flat, it gave an uninterrupted 7ft 1in long double berth sleeping area, a particularly useful feature for outback conditions. The construction of the station wagon increased the weight considerably, and a Zephyr wagon tested by the magazine *Wheels* scaled over 29cwt with just two people on board, and did remarkably well to accelerate from rest to 70mph at this weight in 26.4 seconds, although the maximum speed of 83.4mph was well down on the usual saloon figure. *Wheels* were very impressed with the Zephyr station wagon, and only offered one small criticism, this being that the inside of the tailgate was painted, and so would inevitably quickly become scratched and shabby through luggage being dragged or pushed across it during loading; they suggested that the tailgate should also be covered with the durable floor-covering material. Apparently, the Zephyr was appreciably larger, and therefore more expensive than other station wagons in Australia at that time, and *Wheels* summed it up by saying: 'The Ford Zephyr station wagon is a very nice piece of equipment for the commercial traveller, farmer,

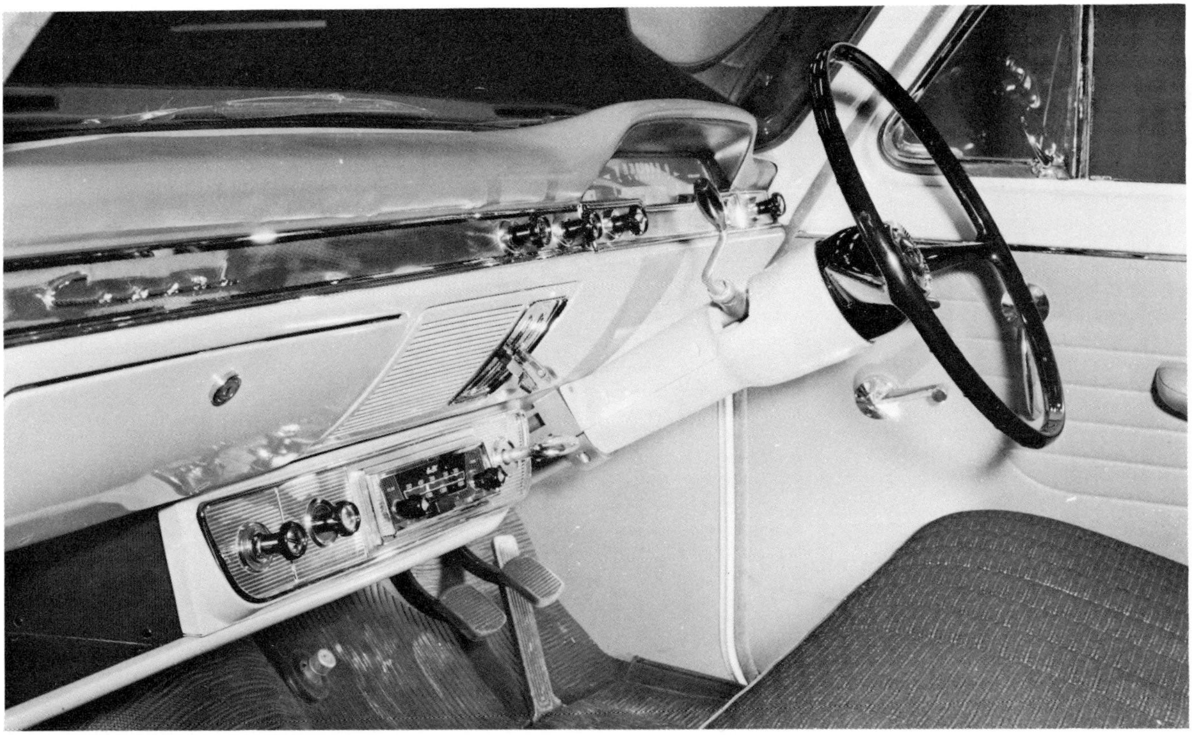

Facia and controls of the 'Low Line' Consul. The speaker for the optional radio was situated behind the grille in the panel above the radio.

shopkeeper, or for the ordinary family man. Rugged, practical and with a good turn of performance, the Zephyr has a slight disadvantage with that £1,430 price tag. Still, the man who wants a bigger than average wagon could hardly look past it'.

A purchase tax reduction in the 1959 budget at home had brought the prices of the Consul, Zephyr and Zodiac down to an attractive £773, £865 and £957, and also helped to boost sales of all makes of car in Britain. Total production of all Mk 2 models during 1959 was a colossal 142,237, and although this peak rate was not maintained throughout 1960, the Mk 2 cars nevertheless continued to completely dominate this sector of the market. Since 1958, works rally Zephyrs had featured front disc brakes in some of the events in which modifications to the competing cars were allowed, and a *Ford News* bulletin on September 28, 1960 announced: 'There are to be no Motor Show changes from Ford this year, but today, the Company announces the introduction of power-assisted front wheel disc brakes as a production option for all versions of the Consul, Zephyr and Zodiac. The extra cost will be £29 15s including tax, making the total price of a disc-equipped Consul £803 – over £180 less than any other disc-equipped British family car. Of special interest to owners of existing Mk 2 models will be the news that Ford Dealers will soon be able to offer a disc brake conversion kit as an accessory at £32 (fitting extra). Tested and approved in the most arduous international rallies, the new Ford disc brakes are equal to all emergencies in the most severe operating conditions'.

At Earls Court in 1960, the centrepiece of the Ford stand was a dual turntable projecting the 'two-car family' theme, featuring a Zodiac and an Anglia, both of which were in a special white finish, with white hide upholstery edged in black and embossed with gold stars, whilst every single

'Low Line' facia and controls, Zodiac and Consul De Luxe. The handbrake was of the twist-release type on these cars, and the instruments included a temperature gauge in place of the previous ammeter.

interior and exterior bright metal part, right down to the tyre valve caps, was finished in 22-carat gold plate. The 542 separate parts in the Zodiac and 306 in the Anglia had been given the 'golden touch' by the Harris Plating Works, of London, and a full quarter-pound of the precious metal had been used. Backing up this glittering display were 10 production Fords, ranging from the 100E Popular to the Zodiac, and no less than five of these cars were awarded medals in the Motor Show Coachwork Competition that year which was organized by the Institute of British Carriage and Automobile Manufacturers. Gold medals were awarded to

Ease of entry and exit was always an excellent feature of Dagenham's big-car range.

The amazing amount of equipment which was swallowed up by the Motorway Patrol Zephyr estates.

the Zodiac, the Consul De Luxe and the Anglia De Luxe. A silver medal went to the basic Consul, and a bronze to the Zephyr, whilst on the Carbodies stand the Zodiac convertible was awarded the silver medal in the Convertible Coachwork category.

Away from the glamour of Earls Court was another Zephyr which certainly deserved a medal. This car was owned by the Goodyear Tyre Company and had spent all its life testing tyres. An account of its amazing record appeared in the *Ford Times*, the report saying: 'As an outstanding example of what performances Goodyear test cars have achieved, a 1958 Ford Zephyr recently completed the astonishing mileage of 417,554 miles in 20 months, using only two engines. The original engine gave 158,000 miles and the replacement engine — still running completely satisfactorily — had totalled 259,554 miles when the car was taken out of the test fleet. As a matter of record, the gearbox was repaired at 217,435 miles; new engine mountings were fitted at 231,871 miles; the back axle had a major overhaul, and new wheel bearings were fitted at 263,792 miles, and the universal joints were replaced at 310,966 miles. This 1958 Ford Zephyr completed the last five months of its test life, running three shifts per 24-hour day on the M1 motorway, at a constant speed of 70mph for one month, followed by four months at a constant 50mph. The car ran on the M1 carrying an all-up weight of 34cwt, a definite overload. Indeed, throughout its running life in the Goodyear test fleet, the Zephyr had never carried less than a 34cwt all-up load, and for the first 12 months of its life on the A5 test route was regularly carrying 38cwt all-up'.

A late technical change was introduced in January 1961. A new gearbox mainshaft, on which the speedometer driving gear was held in position by a nut instead of a circlip, replaced the earlier arrangement, in which it had not been entirely unknown for the circlip in question to break, allowing movement in the gearbox and consequent selection problems. Improved circlips introduced earlier had apparently not been entirely effective in curing the problem, which did not seem to afflict Consuls, but could

The Ford Zephyr estate car.

sometimes occur on high-mileage Zephyrs and Zodiacs where the gearbox was handling the extra torque of the six-cylinder engines.

The Motor put a Zephyr convertible through their road test procedure in the spring for what appears to have been the last road test report on one of the Mk 2 models. Writing in the issue of May 10, 1961, the magazine's staff pointed out that the Mk 2 Zephyr had recently 'quietly celebrated its fifth birthday' and as a result had now 'graduated to the select company of cars which, because they have pleased the customers, sell readily year after year without requiring more than very minor detail improvements'. As far as the performance was concerned, *The Motor* felt that 'for a $2\frac{1}{2}$-litre car to be timed at 90mph is no longer remarkable', but that 'the fine surge of top gear acceleration anywhere in the speed range still makes the Zephyr a brisk moving car on give-and-take roads'. There was praise, too, for the disc brakes with the comment that this option 'may be regarded as adding the finishing touch to a sensibly priced modern version of a high-powered touring car'.

The Motor expressed the belief that the majority of buyers were specifying the disc brake option, and this was indeed the case, to such an extent that in June 1961 this feature was absorbed into the standard specification along with the adoption of sealed-beam headlamps. The Mk 2 models in this final form could be recognized by the deletion of the name script from the upper rear wing and, on the Consul only, by the addition of '375' to its name. The Mk 2 Consul was given the full title Consul 375 at this late stage in its career in order to distinguish it from a recently introduced medium-sized Dagenham product, the full title of which was Consul Classic 315, or, in certain export markets, simply Consul 315. Dagenham had already decided to delete the Consul name from their large car range with the introduction of the appreciably more sophisticated Mk 3 version which was scheduled for the following year. The Consul replacement would be named Zephyr 4 (four-cylinder) with a Zephyr 6 and Zodiac completing the line-up. The Mk 3 range was intended to be considerably more 'up market' than the Mk 1 and Mk 2 models had been, and it was felt at Dagenham that the well respected Consul name would be more appropriate now on a new medium-sized car. Hence the Consul Classic, and the temporary 315/375 identification tags. As the Consul Classic was not in any respect a derivative of the Consul/Zephyr range, and therefore

The space required in which to lower the hood and sidescreens reduces the width of the rear seat on the convertible. The adjustable armrest/doorpull, exclusive to the 'Low Line' Zodiac, can be seen in this view.

Sunshine Express, the 'Low Line' Zodiac convertible.

was not a 'real' Consul, apart from this mention it has no other part at all in the Consul/Zephyr/Zodiac story.

Production of the Mk 2 models was now running at a little more than 1,600 per week, a considerable reduction from the production rates of 1958-60, although very few sales were in fact being lost to rival makes. Many established customers, whilst certainly not in any way 'in the know', had nevertheless correctly reasoned that with the Mk 2 model now well into its sixth year of production, Dagenham must surely be at an advanced

The 'Low Line' Zodiac. Big, bold and beautiful, one of Dagenham's all-time greats.

stage of development with the projected replacement car, and many who had reached this conclusion were now quite prepared to wait and see just what the new model would be. Overseas, too, the sales had taken a downward turn and, through no fault of its own, the Zephyr had now lost its Australian market. A market survey there some time previously had convinced the Ford Motor Company of Australia that a 100% locally built six-cylinder car would soon be needed to satisfy the growing demand, and that the Zephyr would fit the bill. Unfortunately, however, Ford Australia were unable to reach the necessary agreement with Dagenham which would have enabled the Zephyr to be manufactured in Australia. The stumbling block, apparently, was the price Dagenham needed for the Mk 2 Zephyr dies which, of course, they would have had to supply, and so, as the parent Company in the United States were by now producing a Zephyr-sized 'compact' car, the 2.3-litre six-cylinder Falcon, Ford Australia made a successful approach to Dearborn which resulted in the Falcon, rather than the Zephyr, going into production 'down under'.

As production of the Mk 2 models drew to a close, some rather late, but interesting, publicity came their way when the BBC launched a new television programme entitled 'Z Cars'. This featured the escapades of two Police patrol crews who answered to the call signs 'Z-Victor 1' and 'Z-Victor 2'. The two cars were 'Low Line' Mk 2 Zephyrs, and they featured prominently in the early series of this popular programme. These gleaming 'white' Zephyrs were, in reality, painted yellow, as this colour is apparently 'whiter than white' in black-and-white television transmission.

In April 1962, production of 'The Three Graces' came to an end. In just over six years slightly more than 650,000 saloons and convertibles had been produced, with the latter accounting for an exclusive 2% of this total, in addition to which there had been some 30,000 estates, Australian station wagons and coupe utilities. Added to this impressive total, of course, were the preceding 406,000 'Five Star Cars', making a grand total of more than a million Consuls, Zephyrs and Zodiacs in an 11-year period, which remains a production record for this class of car in Britain. Between them they had introduced the comfort and performance characteristics of 'big car' motoring to more people than any other British car before them, and at a new low level of overall economy of operation for this type, which some would insist has never been equalled elsewhere.

CHAPTER 4

Racing and rallying

The Big Fifties Fords in competition

Henry Ford always felt that car racing successes were good publicity, and around the turn of the century he built and successfully raced cars of his own design in an effort to interest potential investors in his ideas. In January 1904, when the Ford Motor Company was only six months old, he announced that he was going to break the world speed record for the flying mile, in order to put the name of his products into the record books and, hopefully, boost the sales of the infant Ford Motor Company. The record attempt was to take place on the frozen Lake St Clair, which was situated just to the north east of Detroit, and cinders from a nearby power station were strewn across the perfectly flat ice surface so that the tyres would have something to grip. On the bitterly cold morning of January 12, at the wheel of one of his 'home made' racers which he had built two years previously, Henry Ford covered the one-mile course in 39 seconds to set a new world speed record for the distance at 91.4mph.

In 1909, a transcontinental endurance race from New York to Seattle, a distance of 4,106 miles, was being widely publicized, and Henry Ford could not resist the temptation to enter the recently introduced Model T. The cars left the starting line in New York on June 1, and 22 days later the Model T crossed the finishing line in Seattle 17 hours ahead of the second-place car, to be greeted by a smiling Henry Ford.

In the years between the two world wars, Fords, or Ford-engined 'specials' of a variety of descriptions, could be seen contesting virtually every kind of motoring event from the Indianapolis 500 to the Monte Carlo Rally. In Britain, club trials were a popular activity for the sporting motorist in the inter-war years, and this kind of competition eventually gave birth to the famous Ford V8-based Allard cars. In the late 1940s, while the planners at Dagenham were working in secrecy on the new Consul and Zephyr Six, the familiar side-valve-engined Fords were back in action at both local and international level. In the hands of Ken Wharton, an export model 10hp Anglia scored outright victory in the 1949 Dutch International Tulip Rally, and the following year Wharton repeated the achievement, this time at the wheel of a V8 Pilot.

With this sort of background it was inevitable that once production was under way the new Consul and Zephyr Six would soon find themselves on the starting grid, and indeed, on the weekend of March 3, 1951, in the North Staffordshire Motor Club's Burnham-on-Sea Rally, a Consul in the

Ken Wharton brings his Consul into the pits at Zandvoort, having just won the 1952 Tulip Rally.

hands of P. J. Collins won the saloon car class in what appears to have been the model's competition debut. Later in the year a Zephyr Six driven by J. R. Smith won the up-to-3,000cc closed car category in the *Daily Express* National Rally.

The entry list for the 1952 Monte Carlo Rally included several Zephyr Sixes, one of which was entered by the experienced Ken Wharton. However, Wharton's Monte came to an abrupt end when the Zephyr left the road and rolled over four-and-a-half times down an embankment before coming to rest upside-down on top of a Citroen, which itself was on top of a Simca. Meanwhile, R. Nelleman, in another Zephyr, kept going well to finish 20th overall in what proved to be a tough event with more than half of the 328 starters failing to finish. Happily, Ken Wharton and his crew were unhurt, and in April Wharton was back in action looking for his third Tulip Rally win. The engine capacity classes and the handicaps imposed on these classes varied considerably from one event to another, and often from year to year, whilst generally being biased in favour of the smaller-engined cars. One of the categories in the 1952 Tulip was up-to-1,600cc and Wharton entered the 1,508cc Consul in this class. The road section of the Tulip was punctuated with special stages on which much higher average speeds were imposed, and additional tests in which bonus marks could be earned and used to offset penalties incurred for lateness on the road or special stages, the final test being a 10-lap race for each class on the Zandvoort Grand Prix circuit. The first of the tests on the 1952 Tulip was a 5-mile mountain hill-climb abounding with hairpin bends, and Ken Wharton took the Consul up in an excellent 8 minutes 26 seconds. Another test, held on the perimeter track at Schipol Airport, involved zig-zagging in and out of pylons in top gear all the way, with an observer sitting beside the driver and holding a rope around the clutch pedal, and so the Consul engine's flexibility was particularly useful here. At the end of the road section the Consul still had a clean sheet plus 18 bonus marks, and needed maximum bonus points in the final race to score outright victory in

The Gatsonides/Worledge car under preparation for the 1953 Monte. The rear seat was removed for the rally to allow the passenger seat to recline into a comfortable sleeping position.

the rally. This entailed being in the lead at the half-way (5 laps) stage and continuing to win. Ken Wharton made a great start, by the half-way stage had the whole of the finishing straight to himself, and when the Consul took the chequered flag to win the beautiful silver Tulip Trophy, it had actually lapped some of the tail-end cars.

Following up Ken Wharton's success, the Ford Motor Company established a small Competitions Department, with E. A. 'Edgy' Fabris as Competitions Manager, and headquarters at the premises of Lincoln Cars (a Dagenham subsidiary which imported American Fords into Britain) on the Great West Road at Brentford, where a small team of mechanics would now prepare works rally entries. Well-known trials driver T. C. 'Cuth'

The Gatsonides/Worledge Zephyr on the final mountain circuit above Monte Carlo.

In an effort to combat the inevitable brake fade which would occur on the final mountain circuit, 'Gatso' arranged for helpers to be in strategic places in the mountains to throw cold water over the brake drums. The light-coloured paint on the front of the car is water-based, and was applied to the rally cars for identification as the mountain circuit roads were open to normal traffic. The two brackets for the anti-glare shield can be seen protruding from the front of the bonnet. 'Gatso' removed the shield for the mountain circuit as conditions were clear and the fog lamps were not needed.

Harrison took a works Zephyr on the Lisbon Rally in November, and although not gaining an award on this occasion, the name of Harrison was nevertheless to become almost synonymous with competition Zephyrs in future years. The experienced Dutch rally driver, Maurice Gatsonides, with eight Monte Carlo Rallies already behind him, including a second place overall in 1950, felt that the Zephyr would be an ideal choice for the tough mountain circuit final stage that so often decided the Monte Carlo winner, and so he approached Ford about the possibility of a works drive in the 1953 event. At this time the Monte regulations did not allow any changes whatsoever in the mechanical specification which would improve the performance, handling or braking, so the modifications incorporated in the works Zephyrs were restricted to those directly affecting crew comfort and convenience. In the interest of weight saving, 'Gatso' decided on just a two-man crew instead of the three which were more normal for the Monte in those days, and so the Zephyr's front bench seat was replaced with bucket seats, the one for the passenger being fully reclining so as to enable Gatsonides and co-driver Peter Worledge to get plenty of sleep and remain fatigue-free on the five-day-long event. A comprehensively-equipped additional instrument panel was mounted centrally on the dashboard to keep the crew informed of engine oil pressure, oil and water temperature and so on. Special snow tyres, supplied by retreading experts Tyresoles, of Wembley, were fitted to the rear wheels. These tyres featured a very bold lugged tread pattern with a central groove around the periphery into which a motorcycle chain could be fitted in a matter of a minute or so if necessary for driving on ice, although, as these tyres were moulded in a softer than standard rubber compound, adhesion on ice proved to be excellent and the chains were not in fact used on the rally. The spare wheel was also equipped with a snow tyre, whereas the front rims were fitted with a pair of

Maurice Gatsonides (left) and Peter Worledge display the trophies which they collected for winning the Monte Carlo Rally. The Zephyr has been washed and the anti-glare shield has been refitted for the presentation.

standard Tyresoles retreads. A large adjustable anti-glare shield, designed by Gatsonides himself, was fitted over the twin fog lamps, and this completed the rally specification.

On Tuesday, January 20, 404 competitors left seven different starting points across Britain, Scandinavia and Europe, and headed for Monte Carlo. 'Gatso' had chosen the Monte Carlo starting point, and the Zephyr

Maurice Gatsonides and Peter Worledge on their return to England with the most famous Mk 1 Zephyr of them all.

Cuth Harrison demonstrates how to cross the finishing line of a special test at maximum speed and still stop short of the penalty line: simply turn the car broadside to scrub off the speed. The event is the 1954 RAC Rally. Photo by The Motor.

first crossed the Alpes Maritimes and the Dauphin Alpes on its way to the Geneva control, from where the route led through Luxembourg, Holland, Belgium and France. Conditions deteriorated after leaving Paris, and the route between St fleur, Le Puy and Valence was covered with old, hard-packed, frozen snow. Commenting on this stage after the rally, co-driver Worledge said: 'Conditions were really bad. It was an ever-twisting road for 150 kilometres and "Gatso" had to drive like mad. Rounding one bend we saw a tractor towing a load of straw over a narrow bridge. The driver turned his tractor into a ditch and the man with him threw himself on top of the load. We shot past but grazed the nearside of the car along the bridge parapet'. The Zephyr arrived at Valence with just 3 minutes to spare, and from there motored on to arrive at Monte Carlo unpenalized. The final mountain circuit stage was 47 miles long and broken into six sections, all of which had to be covered at precisely the same average speed so that time could not be gained on an 'easy' section to compensate for losses on some of the hairpin bend stages, where the timing was almost impossible. In the event, 'Gatso' drove almost to perfection in the mountains, his total time error over the 47 miles being just 2 seconds, and so the Zephyr Six was the winner of the 1953 Monte Carlo Rally. Very close behind in second place was the famed Ian Appleyard in a Mk VII Jaguar, and a little further down the list in 12th place overall was another works Zephyr, in the hands of Cuth Harrison.

In contrast, in the RAC Rally in March, the Zephyrs finished well down the field, and this was followed only a month later by a disastrous Tulip Rally for the Ford team. With just the 10 laps of the Zandvoort circuit to complete, the works Zephyrs were vying with the works Sunbeam-Talbots for the team prize, and Nancy Mitchell was in the lead for the Ladies' Cup in

Cuth Harrison on his way to winning the saloon car category in the 1954 RAC Rally. Many contemporary cars, and indeed some much later models, would lift the inside rear wheel well clear of the ground at cornering forces such as these, and were thus far less controllable than the Zephyr in extreme circumstances. Photo from the Edward Harrison collection.

her Zephyr. However, at the pre-race scrutineering the Zephyrs and the Sunbeams were disqualified on technical grounds as the Zephyrs were fitted with non-standard oil filters, and the Sunbeams with modified brake drums. This was apparently due to nothing more than misinterpretation of the regulations, and the disqualifications were in no way an attempt to deny the British entries the major awards, as in fact they handed the team prize to the Jowett Javelins, one of which was also the outright winner of the rally.

The regulations for the 1954 Monte allowed modified engines for the first time in this event, although additional handicapping was imposed on those cars so equipped. The works Zephyrs were entered in the modified category, running with a high-compression cylinder head and triple SU carburettors. The routes from all starting points finally converged on Valence, from where the competitors continued to the control at Gap. The final 164 miles from Gap to Monte Carlo formed a tightly timed regularity test, the most difficult section being the ascent and descent of the loose gravel-surfaced Col des Leques. The Zephyrs were apparently the most spectacular cars on the Col, according to *The Autocar's* report: 'The Zephyrs particularly travelled faster than any ordinary motorist would dare attempt. T. C. Harrison was typical of this Zephyr contingent, going into a corner just after the summit at what seemed an impossible speed. However, he was checked by a Daimler being driven more slowly and must have been almost blinded by the dust. Sydney Henson's Zephyr was also baulked, this time by a Sunbeam-Talbot'. The regularity section was in place of the previous year's mountain circuit, and the final stage at Monte Carlo this year was five timed laps round the Monaco Grand Prix circuit for the highest placed 100 cars. Seven Zephyrs, including some private entries, were in the top 100, the best placed being the works-entered cars of Jack Reece (14th) and Cuth Harrison (15th). In the speed test Harrison put up the fifth fastest lap of the 100 competitors, and this performance lifted him to 13th overall in the final reckoning. For the RAC British Rally in

Edward Harrison (inside) and Fotheringham-Parker have a private duel at Zandvoort in the 1954 Tulip Rally. VHK 196 was a veteran competitor by this time, having already contested two Montes, two RACs and the previous year's Tulip. Soon after this, it completed the Liège-Rome-Liège in the hands of Nancy Mitchell. Photo from the Edward Harrison collection.

March, Cuth Harrison was at the wheel of the same Zephyr, XNO 496, although it was now with a standard cylinder head and single carburettor, as modified cars were not eligible for this event in 1954. Harrison put up an excellent show, winning the saloon car category outright and gaining third place overall behind two Triumph TR2s. The Zephyr was also a member of the winning team, sharing this award with the works-entered Anglias of Jack Reece and Nancy Mitchell.

Both Consuls and Zephyrs featured in the entry list of the 1954 Dutch International Tulip Rally. Cuth Harrison, once again in XNO 496, with his son Edward as co-driver, built up an early lead in the up-to-2,600cc class, whilst a Jowett Javelin held the two Consuls of Jack Reece and W. Fleetwood at bay in the 1,600cc category. By the half-way stage Harrison was in fourth place overall behind an Alfa Romeo, a Porsche and a Jaguar, when he 'lost' the Zephyr on an apparently unexpected icy patch in the Ardennes and the car ended up in a snowbank. The Zephyr was undamaged, but the time lost was sufficient to put the Harrison duo out of the running for major awards, and the renowned Sheila Van Damm now led the 2,600cc class in her works Sunbeam-Talbot. By the end of the road section, the two Consuls had pulled ahead of the Javelin and in fact were now so far in front in this class that their 10-lap race at Zandvoort was just a formality to see which of them would win the class — in the event it was Jack Reece. Reg Philips had brought his Zephyr up to share the 2,600cc class lead with Sheila Van Damm, although with the cars running in standard form it was generally thought that the faster Sunbeam-Talbot would beat the Zephyr at Zandvoort. Philips handed over the Zephyr to his co-driver Denis Scott for the all important race. As expected, Sheila Van Damm took an early lead, but what the Sunbeam could gain on the straights the Zephyr could pull back in the turns, and Scott was giving a masterly display of the Zephyr's superb handling qualities, sometimes bringing the Zephyr within striking distance of the faster car out in front. Then, right at the end of the 26-mile race, almost incredibly it emerged that Scott had been keeping something in reserve, and he brought the Zephyr

Cuth and Edward Harrison with their wire-wheel Zephyr on the 1954 Alpine Rally. Photo from the Edward Harrison collection.

through beautifully on the last turn to take the chequered flag literally by inches, to win the class for Ford. Denis Scott's display completed a brilliant hat-trick for Ford in this event, for in addition to Jack Reece's class win in the Consul, Maurice Gatsonides had taken the 1,300cc GT category with a twin-carburettor Anglia.

Unfortunately, the Ford works cars were unable to contest every major event at this time, as the Competitions Department was running on an extremely tight budget. Whilst competition successes certainly added

Jack Reiss swings his Zephyr Six round a pylon in the final test on the 1954 RAC Morecambe Rally. RNW 2 featured a fully-modified engine, with high-compression cylinder head and single-piece copper gasket, double valve springs, triple SU carburettors, and six-branch exhaust system. It was a regular sight at northern rallies, races and sprints for a number of years during the 1950s. Photo from the Jack Reiss collection.

prestige to the Company's name, these same successes could not result in an increase in overall sales at a time when Ford agents everywhere had waiting lists of customers for every model in the range, and so Dagenham simply could not justify a large expenditure on ventures which, even though successful, meant no financial gain whatsoever. This situation certainly did not make things easy for Edgy Fabris and his team at Lincoln Cars, or for the works drivers, who were not always getting the back-up which some other competitors enjoyed. The results gained so far were all the more creditable when related to the limited finances available.

One important event which had not been contested so far by the works team was the famous Alpine Rally, although motoring journalist Bill McKenzie had done well with a Zephyr in the early stages of this event in 1952, before retiring following a crash as a result of braking problems. Cuth Harrison borrowed a Zephyr, XNO 496 again, from the Competitions Department and entered it privately for the 1954 Alpine. One of the biggest problems facing drivers in this event was the almost inevitable brake fade which occurred as a result of the succession of high-speed descents of Alpine passes, and Cuth Harrison sought to overcome this problem by converting the hubs on XNO 496 to take wire wheels, which would allow more cooling air around the brake drums. However, the wheels chosen did not stand up well to the violent cornering tactics employed on the rally by Cuth and Edward Harrison, and with broken spokes becoming a problem the Zephyr was pulled out of the event in the Dolomites.

In August, Nancy Mitchell borrowed another works Zephyr and entered it privately in the tough Liège-Rome-Liège. There were only four British entries in the 99-strong field, and the almost impossible schedule laid down managed to reduce the number of finishers to just 34, one of which was the lone Zephyr in 28th position. Whilst Nancy Mitchell was pressing the Zephyr hard across Europe, another Zephyr put up a remarkable performance in the 12-hour American endurance race held at Linden New Jersey Airport on August 22. The Zephyr, a 1952 model purchased secondhand by Bob Bathhurst only a few weeks before the race, was the only Ford in the field of 39, which was made up of an assortment including 18 Jaguars, several Austin-Healeys, Porsches and MGs and an Aston Martin. The Zephyr ran trouble-free, making only five pit stops for fuel and oil/water checks during the 12-hour stint. Only 20 cars lasted the 12 hours, and at the end the Zephyr was placed fourth overall behind three surviving Jaguars.

The list of starters for the 1955 Monte revealed that 22 of the big Fords were bidding for honours, 19 of which were the six-cylinder Zephyrs and Zodiacs. A strong works contingent was included in this number with the now familiar names of Cuth Harrison, Denis Scott, Jack Reece and Reg Philips in modified Zephyrs, and the well-known competition driver Sydney Allard with a standard Zodiac. BBC commentator Raymond Baxter was covering the rally from Reg Philips' Zephyr. Notable amongst the private entries was G. N. 'Gerry' Burgess. In the previous year's event he had participated as co-driver in J. Risk's Zephyr, and had been suitably impressed with both the event and the excellent performance of the car. Soon after this he purchased a brand new Zephyr himself, and it was this car which he entered for the 1955 rally, with the experienced Ian Walker as his co-driver. The Zephyr was running in the standard category, and as a private entry did not qualify for any works support. Preparation was limited to little more than a thorough routine-type service so as to ensure that the car was in optimum standard condition. Two spare wheels, shod with snow tyres, were carried in addition to the car's four Michelin Xs which Gerry Burgess favoured for normal use. At almost the very last moment, co-driver Ian Walker was forced to withdraw and he suggested Peter Easton as substitute. Burgess had chosen Monte Carlo as his starting point, and on the way to Dover he picked up Peter Easton, whom he was meeting for the first time. Road conditions in Kent were bad, with snow falling in blizzard proportions, and abandoned vehicles were causing serious obstructions in many parts. Easton took over in order to familiarize himself with the Zephyr under adverse conditions and was desperately unlucky to be confronted with an unlit vehicle which had been abandoned by its driver. In the collision that ensued the Zephyr sustained extensive front end damage and was in no condition to continue to the start. The Ford Agents, Haynes of Ashford, did a wonderful 'instant' front-end repair job, and later the next day the Zephyr and its crew were once more on the way to Dover, and then to Monte Carlo for the start.

As the rally got under way, bad weather conditions in Britain and across Europe ensured that whichever starting point had been chosen, no competitor would have an easy run. Secret checks along the route were an added hazard, and as the cars began to converge on Monte Carlo at the end of the 2,000-mile-long road section not one had escaped incurring penalty points. Cars in the 'modified' category had to make up an 8% penalty over their standard counterparts on the timed sections, and this had generally proved to be too much to ask on the hard-packed snow and ice which had prevailed over much of the route. Consequently, it was the standard models which were occupying the leading positions at this stage. The Burgess/Easton Zephyr had had a good run and on arrival at Monte

Gerry Burgess on the final five-lap speed test at Monte Carlo in 1955. Photo by The Motor.

Carlo Gerry Burgess found himself leading the rally. The 100 highest placed cars had still to contest the mountain circuit, and those who survived the gruelling 180-mile course which had been chosen this year would then have to complete a five-lap timed race on the Monaco Grand Prix circuit before the overall results could be decided. An unsuspected secret check had been placed on the road section south of Grenoble at the end of a long fast stretch and many competitors had been caught out here, including the Aston Martin of Monte Carlo veteran Maurice Gatsonides and the green Ford Zephyr of Gerry Burgess. An alleged 7 minutes early at this check added 70 penalty points to the Zephyr's total, and Gerry Burgess now found himself back in 10th position. Conditions for the mountain test were atrocious, hard-packed snow covered virtually the entire 180 miles, sleet and snow was falling in the hills, and the clouds were clinging to the mountain sides, giving foggy conditions along many sections of the route.

The 100 qualifying cars left Monte Carlo at 1-minute intervals and headed into the mountains. Burgess pressed on at an incredible pace, passing many of the competitors who had started out from Monte Carlo ahead of the Zephyr before reaching the Col de Turini. On the twisting climb to the summit of the Col, after more than 100 miles of punishment, the Zephyr's brake fluid boiled and the pedal sank to the floor. Until the brakes had cooled, slowing down for the hairpin bends on the long descent was accomplished by slamming the Zephyr into first gear and putting it sideways on the hard-packed snow to scrub off excess speed. The brakes eventually came back, but the punishment taken by the gearbox in the meantime had now resulted in the loss of top gear completely. Nevertheless, Burgess pressed on regardless, screaming the Zephyr along in second gear, and by the time it swept back into Monte Carlo it had climbed from 10th to fourth place overall, thanks to a magnificent drive. The next best placed Ford was the Zodiac of Sydney Allard, now in 13th place overall after another spectacular mountain drive which had been such as to excite *The Autocar* reporter to comment: 'The prize for the most daring descent might well go to Sydney Allard, who slid his Zodiac around

Best of British. Gerry Burgess (left) and Peter Easton display the Stuart Trophy which was awarded to the highest placed British car/crew combination at Monte Carlo. Their privately-entered Zephyr had finished fourth overall on this, the 1955 event. Photo by The Motor.

Cuth Harrison and son Edward after winning the up-to-2,600cc GT category in the 1955 RAC Rally. The bonnet top of the Zephyr has been adjusted at its rear edge to allow warm air from the engine compartment to flow over the screen. Photo from the Edward Harrison collection.

with fantastic abandon'. All that now remained were the five-lap races around the Grand Prix circuit, and unless someone failed to finish their particular race it was unlikely that any drivers could improve on the positions they now held. Leading the rally was Norwegian Per Malling in a Sunbeam; he, too, had put up an excellent display in the mountains, having climbed from fourth to first, but at the expense of the Sunbeam's engine, which was now described as 'distinctly rough', and doubt was being expressed in some quarters as to whether it would last out its five-lap race. In the event it did, just, Malling actually being lapped at the end, but

Ralph Sleigh pushes his Zodiac hard through a climbing turn on the ascent of the Col de La Faucille in the 1955 Tulip Rally. Photo by The Motor.

Vic Preston and D. P. Marwaha after winning outright the East African Safari in 1955. This win, coupled with Gatsonides' 1953 Monte Carlo success, gave the Mk 1 Zephyr a rallying double which it shares only with the Mercedes 220.

finishing at all was sufficient for him to hang on to his hard-earned lead.

Gerry Burgess removed the snow tyres from the 'two-speed' Zephyr and put it back on Michelin Xs all round, inflated to 48psi for the 'round the houses' race. He recalled recently: 'I was due to run in one of the last races, so that I knew before I started that I had to achieve a lap time 4 seconds faster than the best of the Ford team had done. I did it, in second gear, and the recording of Raymond Baxter's commentary where he refers to "Gerry's 10,000-revs Ford" has always been a favourite memory'.

Real winter weather conditions in March gave the entrants in the 1955 RAC Rally a tough time throughout the event, and on the first night, in Wales, more than a fifth of the field retired. The works Zephyrs kept going well, although Sydney Allard hit a wall in Wales, and later a bridge parapet in the Lake District, which did extensive damage to the Zephyr. Cuth and Edward Harrison had a good run all the way in spite of the bad conditions, and drove their Zephyr to a class win. Allard kept the damaged Zephyr going to take second whilst Jack Reece came in third to complete a Zephyr hat-trick in the 2,600cc GT saloon category.

In the Tulip Rally, in May, Cuth Harrison took yet another class win, this time with a Zephyr equipped with a Laycock overdrive. According to the regulations for the event the overdrive was considered to be a modification, and the Fords so equipped were given an additional handicap by being placed in the up-to-3,500cc category. The handicap for this class was the same as for some of the sports classes, and for the final test at Zandvoort Cuth Harrison found himself sharing the starting grid with cars like Austin-Healeys and Triumph TR2s. Nevertheless, at the end of the 26-mile race the Zephyr crossed the line in third place, only 15 seconds behind the winner and $2\frac{1}{2}$ seconds behind the second place car, both Triumph TR2s, a performance which earned Cuth Harrison a well-deserved standing ovation from the grandstand crowd. Earlier, Sydney Allard, in a standard Zodiac, had also given the Zandvoort crowd something to cheer about in his race when he had lapped the entire field except for another Zodiac driven by Ralph Sleigh, although as both these cars had incurred many penalties on the earlier stages they were out of the running for awards.

In the *Daily Express* Trophy meeting at Silverstone, in 1955, one of the races was for Production Touring Cars. All the entries in this 25-lap (75-miles) event had modified engines, and the lone Zephyr in the field, although being driven by Ken Wharton, was in fact owned by none other than Raymond Mays. Two Consuls, one modified by Laystall and the other by Derrington, were also amongst the 27 starters. The C Type-engined Mk VII Jaguars inevitably took an early lead, whilst Ken Wharton found himself in the midst of a tangle of Riley Pathfinders. Unfortunately the Consuls fared badly, the Derrington car going out early on with a burst radiator hose and the Laystall car finishing well down amongst the tailenders. The Zephyr finally shook off the Pathfinders on lap 5 and Ken Wharton then settled down to finish comfortably in fourth place behind the three Jaguars, with the Zephyr's average speed for the race of 76.18mph comparing very well against Mike Hawthorn's 78.92mph in the winning Jaguar.

In sharp contrast to the major European rallies was the East African Safari. In the 1955 event, 57 strictly standard cars left the start in Nairobi to follow a 2,490-mile route which would take them across Kenya, Uganda and Tanganyika before eventually leading back to Nairobi, along rocky dirt 'roads' which, according to the *Mombasa Times*, 'would deter a tank'. On the first stage, a Zephyr driven by Vic Preston and D. P. Marwaha hit a vulture and as a consequence suffered a broken windscreen, but the crew pressed on and later on, with some time in hand, they had a replacement screen fitted. After the first 1,000 miles Preston's Zephyr was one of the few cars still unpenalized, but after that, conditions worsened considerably and only 20 cars made it to the finish in Nairobi. Three Zephyrs were amongst the surviving number, with one driven by Mary Wright winning the Lady McMillan Memorial Trophy (Ladies' prize) and that of Vic Preston being the outright winner of the rally. Vic Preston's Safari, coupled with Gatsonides' Monte Carlo and Ken Wharton's Tulip, had given the Mk 1

Ken Wharton at Silverstone in Raymond Mays' Zephyr in 1955.

Cuth Harrison on his way to 18th place overall on the 1956 Monte Carlo Rally. The horse-shoe on the grille was making the first of a succession of appearances on a Harrison rally car and is today amongst Edward Harrison's mementoes. Photos from the Edward Harrison collection.

model a hat-trick of outright wins at the highest level in three widely differing events, which served more than anything else to underline the remarkable competence and versatility which it had shown as a competition car.

The new Mk 2 Zephyr made its international rallying debut in the Alpine Rally in 1956. As its title suggests, the Alpine is almost wholly a mountain rally, consisting of ascents and descents of numerous mountain passes at an 'impossible' schedule which was set so as to virtually ensure that no competitor would be unpenalized at the end of the road section (on top of penalties incurred on the special stages). Those who did occasionally manage to beat the road schedule and finish unpenalized were awarded a Coupe des Alpes (Alpine Cup) irrespective of their final position as decided on the special stages. Modified engines were allowed in each category in the 1956 event and the three works Zephyrs were fitted with triple Zenith carburettors. The new Mk 2 cars went so well that Edward

Still going strong; Jack Reiss' Zephyr was probably the last Mk 1 model to compete in national events, and is seen here in action in 1958, by now with more than 120,000 miles behind it, and still on the original engine. Photo from the Jack Reiss collection.

The Mk 2 Zephyrs display their trophies at the end of the 1956 Alpine Rally, their first event. Left to right are Anne Hall and Miss Jackson, Cuth and Edward Harrison, Denis Scott and Stan Astbury. Photo from the Denis Scott collection.

Harrison, who was co-driving once again for his father, recalled recently that the only problem which occurred was when he suddenly thought he had left the road book at the previous control point; the Zephyr was turned round and headed back to the control when the book was found behind the sun visor. Cuth Harrison brought his Zephyr in to win the up-to-2,600cc class, with Denis Scott's similar car taking second place, and as both had clean sheets on the road the Mk 2 Zephyr had won two Alpine Cups at its very first attempt. Driving the third Zephyr in the works team in this event was Mrs Anne Hall. After co-driving for Sheila Van Damm in the works Sunbeams for the previous four years, Anne Hall had found herself suddenly without a works drive after the retirement of Miss Van Damm and so had moved over to Ford. Although just failing to gain a Coupe des Alpes, her first Zephyr drive was nevertheless very impressive, and she had motored up the snaking Izoard Pass special stage faster than any other competitor in the rally, a performance which earned her a special trophy and the unofficial title 'Queen of the Izoard'. Anne Hall still remembers another incident on that particular Alpine. At one of the overnight stops, as the crews were having supper, she mentioned that her hands were badly blistered. Cuth Harrison suggested that as the Zephyr was appreciably larger and heavier than the Sunbeam perhaps she ought to wear driving gloves until becoming more familiar with the new car. The following morning, before the dawn start, Anne Hall found a pair of new, soft leather gloves at her place at the breakfast table. After the others had retired the night before, Cuth Harrison had been out to the village and persuaded a shopkeeper to open up and sell him a pair of gloves.

The fuel shortages caused by the Suez crisis led to the cancellation of several major rallies in 1957, the Monte Carlo, RAC and Alpine all being victims. The Tulip Rally was unaffected, and a strong works Zephyr contingent was amongst the starters with Gerry Burgess having his first works drive. Unfortunately Burgess went out with a fuel starvation problem, but the remainder of the Zephyrs had a trouble-free run with Anne Hall winning the Ladies' Cup and the three cars of Denis Scott, Cuth Harrison and Ronnie Adams combining to win the Team Trophy, which was normally awarded to the three cars of the same model with the lowest aggregate penalties.

In August, Anne Hall took a works Zephyr on the Liège-Rome-Liège, although as a private entry without works support, and Gerry Burgess

'The Queen of the Izoard'; Anne Hall at the wheel of an early Mk 2 Zephyr. Photo from the Anne Hall collection.

entered his own Mk 2 Zephyr, with an 11:1 compression ratio engine specially prepared by Raymond Mays for this almost racing rally. All went well on the first stage, but after leaving Munich, Anne Hall's Zephyr developed wheel bearing trouble, and with no support at the control points this meant leaving the route to find a suitable garage. Almost an hour was lost having this trouble rectified, but with some spectacularly fast driving, Anne Hall made up all but 3 minutes of the loss by the next control. After over 1,300 miles were completed her Zephyr was hit hard by an Army lorry in northern Italy, the driver of which had totally ignored the car's priority position on a narrow bridge, and more time was lost lashing together the stoved-in nearside doors with the tow rope. Despite this damage, and the

Anne Hall's Zephyr on the 1957 Liège-Rome-Liège, soon after being hit by an Italian army lorry. Photo from the Anne Hall collection.

fact that the collision had left the exhaust system trailing on the road, Anne Hall took the Zephyr up the 6,900ft Passo de Giau special stage in the third fastest time of any competitor. But making up so much time after these setbacks had punished the brakes so severely that the linings were well and truly cooked, and the brakes simply would not recover. Before there would be any opportunity to have the brake shoes replaced Anne Hall was faced with the descent of the Stelvio Pass. She almost made it, negotiating more than 30 hairpin bends using only the gearbox as a brake before her luck finally ran out on the third hairpin from the bottom, where the Zephyr hit the retaining wall, so bringing a magnificent drive to an end. Anne Hall's performance so impressed John Gott, the BMC team captain on the Liège, that he wrote to the Ford Motor Company praising her effort, and concluded: 'I do feel, however, and this you may feel impertinent, that it was not really fair to send out a lady driver of this class without some support. The male drivers had been able to change their brake shoes, but Anne had not and it was trying to make up time to get into a garage that was one of the factors resulting in her crash.'

Meanwhile, Gerry Burgess' Mays-tuned Zephyr had no trouble at all keeping up with the leaders, but unfortunately, after his spare supplies of 100-octane fuel had run out in Jugoslavia, he had been forced to fill up with the local 'brew', which at an 80-octane rating was totally unsuitable for the 11:1 compression ratio engine. Eventually, as a result of the sustained detonation which had inevitably taken place, the piston crowns began to break up, and the Zephyr came to rest only 150 miles from the finish. Close behind was another competitor, Schlesser, in a Mercedes 300SL, and as he had some time in hand he very sportingly towed the stricken Zephyr to a

Facia panel of an early Mk 2 Zephyr rally car. In front of the navigator are two stopwatches, whilst the oblong instrument with the two faces, mounted next to the steering column, is the Halda Speed Pilot, which, once set with the distance to be covered and the time allowed, informed the crew whether they were ahead of, or behind time at any stage on the route. Photo from the Norman Masters collection.

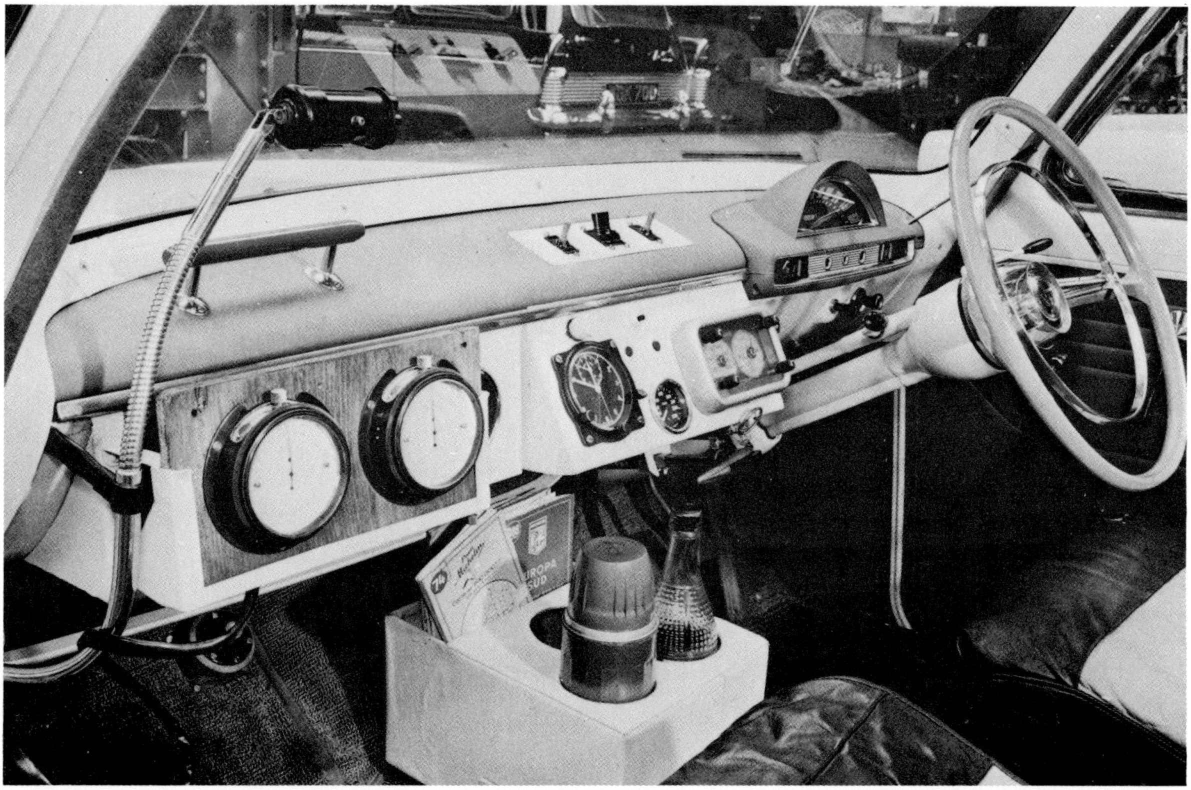

nearby garage before continuing on his way to take second overall in his privately-entered Mercedes. Bits of broken piston were retrieved through three of the plug holes, and the Zephyr was then push-started down a nearby slope. On the three remaining cylinders it pulled quite well, and Gerry Burgess pressed on towards the finish. On the descent overlooking the finishing point a fourth piston broke up and seized the engine, but it was downhill all the way and the Zephyr coasted over the line in 12th place overall, the highest British finisher.

Seven works Zephyrs contested the 1958 Monte Carlo Rally, three of them running with modified engines featuring triple carburettors and cylinder capacities increased to 2,648cc by reboring 60-thou oversize. On the night of January 21/22, 303 starters left the eight different starting points and headed for Monte Carlo and into conditions which at the end were generally reckoned to have been the worst ever for this event. As usual the Monte attracted many private entrants, one of which was Eric Jackson, from Barnsley, starting out from Glasgow in a recently acquired secondhand Mk 2 Zephyr. The first of the Glasgow contingent left at 3.40am, and as the Jackson/Boardman Zephyr made its way out of Stranraer at dawn the gearbox failed. By sheer luck the failure occurred near a Ford dealership, and only two hours after waking up the proprietor, the secondhand Zephyr was heading south with a new gearbox 'borrowed' from a Zodiac in the dealer's showroom. As anything over an hour late at the major control points meant automatic exclusion on the Monte, Jackson had a hectic drive south in the wintry conditions and succeeded in pulling back an hour or so of the time lost, clocking in at Dover 59 minutes late, just 1 minute inside exclusion time.

The 85 Paris starters got away in clear weather, but soon met with adverse conditions. Snow began to fall with such intensity that at times it was like a white fog, and competitors were soon dropping out at an amazing rate. Despite the conditions, Denis Scott made good time with his modified Zephyr in the early stages. Running on Pallas snow tyres, Scott was able to keep the Zephyr going well enough to come up really close behind slower moving competitors who were holding him up, and so intimidate them into going too fast into turns and consequently leave a gap which he would then take the Zephyr through. 'We worked our way through an awful lot of vehicles in this way', said his navigator Ken Armstrong recently. However, a lorry completely blocking the road brought their rally to a premature end. Further in front, Edward Harrison was making good progress in another of the modified Zephyrs, and talking about the event recently, he said: 'I remember at one point I didn't leave the driving seat for nearly 20 hours during the drive through the Jura mountains and the Massif Central. We saw an awful lot of cars upside-down, on their sides, in ditches and fields; and then we seemed to motor on for quite a long time without seeing another car, which was quite disconcerting. At one point we put one of the spare wheels with snow chains on the offside front to help steer in the slippery conditions. This was OK until we got on to some tarmac and the friction heated the chain up, which then burnt into the tyre. Fortunately we discovered this before it went right through. I would, of course, like to pay tribute to my crew — Dick Habershon and Jim Furse. How their nerves stood up to it I will never know.' Edward Harrison and his crew made it to Monte Carlo, their Zephyr being the only one of the 85 Paris starters to get through.

The privately entered Jackson/Boardman Zephyr was also amongst the 59 finishers, after managing to pull back more time across Europe and

Denis Scott drifts an early Mk 2 Zephyr at speed during a production car race. The Zephyr is running on Michelin X tyres, inflated in this instance to around 60psi. Photo from the Denis Scott collection.

build up a safer margin than the slender minute with which it had left Boulogne — a particularly good effort, this, in conditions which finally eliminated more than 75% of the entry. The highest placed of the four surviving big Fords was another private entry, a Zodiac which had started out from Rome in the hands of R. Nelleman and M. Skarring, who brought it in to win the over-2,000cc Touring Car class and gain 10th place overall.

In March, Arctic conditions in Britain embraced almost all of the 1,800-mile route chosen for the RAC Rally. Amongst the Ford entries, Denis Scott and Ken Armstrong were in the modified Zephyr which they had used in the Monte, and on the second night, in Wales, they were still unpenalized when the Zephyr came to rest, having run out of petrol as the result of a most unfortunate miscalculation. Nevertheless, when the starters left Blackpool on the Wednesday night for the snowbound northern section of the rally, Scott was still well placed in the GT class despite the time lost in Wales. The snow was thick in the north, but with Firestone Town and Country tyres on the rear, and a set of chains for the really difficult sections, the Scott/Armstrong Zephyr continued to make good time. Then, coming over the brow of a hill at about 80mph at the end of a long straight stretch of road in Cumberland, Denis Scott was confronted with a bend which was far tighter and much nearer the brow than it appeared on the map. There was no way the Zephyr would get round, and at the turn it left the road, ran up the verge and became airborne. The ground fell away after the verge, and the Zephyr went through the top of a hedge and completely cleared an adjacent fence before landing all square in a snow-covered field. 'It was just like Becher's Brook' said Scott when recalling the incident. Further along the fence over which the Zephyr had just flown was a gateway on to the road, and so within minutes Scott was back on course. Only a mile or so further along, the temperature gauge needle was off the scale, and Ken Armstrong guessed correctly that the radiator tap must have been turned open as the car went through the top of the hedge. Attempts to replenish the cooling system with snow only resulted in huge clouds of steam; however, the provisions on board included several bottles of Lucozade, and with these emptied into the radiator the Zephyr was back in action. With no more

Denis Scott negotiates the final test on the 1958 RAC Rally. Photo from the Ken Armstrong collection.

'excursions' — 'I only allowed him one excursion per rally' says Ken Armstrong — Denis Scott brought the Zephyr in to the finish to win the over-2,000cc GT category ahead of Lyndon Sims' Aston Martin and B. Mitton's 3.4 Jaguar.

The major manufacturers did not yet officially contest the East African Safari, the line-up being made of locally entered cars, some of which were sponsored by a garage dealing in the make concerned, but with many being simply private entries. The 1958 event was split into three categories — Impala, Leopard and Lion class — which were decided on price, but nevertheless came out conveniently as small, medium and large. Several Zephyrs were in the Lion class, with the principal opposition coming from

Denis Scott and Ken Armstrong at the end of what had been, for them, a very eventful RAC Rally in 1958. Their Zephyr was the winner of the over-2,000cc GT category. Photo from the Ken Armstrong collection.

the Mercedes 219/220 series and a team of Jaguars. At the end of the 2,900-mile event, the leading Zephyr, crewed by the Kopperud brothers, had lost just 3 minutes on its schedule and so incurred 150 penalty points, which appeared to give it the joint outright win with the leading Leopard class car, a German DKW, also with 150 penalties. The Zephyrs were first, third, fifth and sixth in their class, with Mercedes occupying second and fourth positions, and with the three leading Zephyrs' total penalties of 6,050 being less than any other three-car combination the Team Prize also looked secure according to the provisional results. However, when the final results were posted, the leading Zephyr was shown as second in the Lion class, with an additional 1,000 penalty points handed out at the scrutineering for a broken shock absorber seal. The Impala class-winning Anglia had been given an extra 2,000 penalties for the same reason, putting it behind four Volkswagens, three of which were awarded the team prize, apparently because of their now 1,2,3 position in the Impala class, despite the fact that their total penalties of 7,850 were still higher than the Zephyrs' 'new' total of 7,050. So, what at first had looked to be a Ford-dominated event suddenly became an almost wholly German affair, although Mrs Mary Wright's Ladies' Cup-winning performance in her Zephyr was confirmed. The seals at the centre of the disputed penalties were those of the wire and lead variety fixed to certain components by the organizers at the beginning of the rally, and some of these had simply worked loose or been broken by the pounding on the 3,000-mile route. As can be imagined, protests followed, and since it was obvious that the Kopperuds' Zephyr had not had any components changed it was reinstated as the Lion class winner.

The Zephyr's previous Tulip successes were not repeated in the 1958 event. The three cars of Denis Scott, Cuth Harrison (now with younger son John as co-driver since Edward had graduated to being a works driver in his own right) and Edward Harrison, vied with each other for the 2,600cc class lead in the early stages before both the Harrisons went out with fuel starvation. Although foreign matter was found in the petrol in both cases it appears to have been sheer coincidence that it had afflicted just the Harrison-manned cars. The Scott/Armstrong car lost the class lead when it

Jeff Uren battles with Doc Grace's Riley Pathfinder in 1958. Photo from the Jeff Uren collection.

followed another competitor's tail-lights round a wrong turning on a foggy mountain climb and was late at the next control.This mistake handed the 2,600cc class to the Morley brothers in a 2.4-litre Jaguar. Although being held in late April, the Tulip route still usually featured snow and ice in the hilly regions, and it was normal on this event for the Zephyrs to take two spare wheels shod with snow tyres, usually the Firestone Town and Country cross-ply type. As the other four tyres were almost always the radial-ply Michelin X type, this resulted in the Zephyrs often running on a front radial/rear cross-ply combination, a mix which is illegal in Britain today and yet which apparently worked well on the Zephyr. Denis Scott won his race at Zandvoort at the end of the 1958 Tulip on this combination, as one of his X tyres had been punctured on the road section.

Three of the six Zephyrs amongst the 58 starters in the Alpine Rally in July had disc brakes on the front wheels for the first time. In the event, however, it was the standard Zephyr of Edward Harrison which put up the best Ford display in what proved to be the toughest Alpine Rally yet, with only 25 of the original 58 who set out actually making it to the finish. Edward Harrison finished fifth overall behind three Alfa Romeos and a Triumph TR3, and just ahead of a shoal of Austin-Healey 100s.This class-winning performance also added the third Coupe des Alpes to the Zephyr's total.

Although already very familiar with the Mk 2 Zephyr as a racing saloon, Jeff Uren had his first works Zephyr drive in the 1959 Monte, and brought the car in 13th overall at the finish, the highest placed of the big Fords at Monte Carlo on this occasion. In March that year, Denis Scott and Edward Harrison, accompanied by Competitions Manager Edgy Fabris and mechanic Jack Welch, flew to Nairobi as the first-ever works contingent in the East African Safari Rally. One of the three Zephyrs entered by the works was to be crewed by local drivers, whilst Scott and Harrison would each team up with a local co-driver whose knowledge would be an essential asset at this stage as no suitable maps of the area existed. The 63 entrants left Nairobi on Good Friday and headed out on the southern loop. The Scott/Davies car first hit trouble approaching Mbulu, after the Zephyr had been bottoming almost continuously on a tight section with a 41mph average; when Scott braked for a turn the pedal just sank to the floor. The Zephyr made it round after hitting the bank on the outside of the corner and was pulled up on the handbrake. One of the brake pipes had been knocked loose, but fortunately not broken, although some time was now lost refilling and bleeding the system. The next section was 46 miles of

Jeff Uren holds a slight lead over Jack Sears at Brands Hatch in 1958 in one of several dices the pair had that year. More than 20 years later, Jeff Uren still rates the Jack Sears/Austin Westminster combination as one of the finest ever in saloon car racing. Photo from the Jeff Uren collection.

A considerable drop over the edge of the road does not stop Denis Scott from hanging the tail out as he hurries to Monte Carlo in 1959. Photo from the Denis Scott collection.

unmade track at a 42mph average, and all went well until 7 miles from the end when, just after overtaking a slower-moving competitor, the Zephyr went into a huge pothole. Denis Scott remembers: 'I thought we would turn completely over, but the car somehow bounced out. During the quick stop to see what damage had been done the car we had just passed came by, the crew laughing their heads off; however, 5 miles further on we had the last laugh as we came across them with the sump of their car firmly embedded on a large rock.' A bent track rod was the only apparent damage, and the Zephyr arrived at Dabil exactly on schedule. The 'road' improved slightly between Dabil and Banga, 37 miles away, and the average was stepped up to 50mph. Even so, some time was made up, and with a few precious minutes to spare the Zephyr was stopped in Banga for a quick check-over by the crew and had the fuel tank replenished from the jerrycans on board. The 50mph average continued for the next 38 miles, but conditions worsened and the Scott/Davies Zephyr was a minute late at the Kolo control. Denis Scott put in a terrific effort over the next 110-mile stage to Dodoma, gaining enough time to take the Zephyr into the Ford agents there and have the brakes properly bled and the track readjusted in an effort to lessen the excessive tyre wear being caused by the bent track rod. By the time Scott reached Nairobi and the end of the southern loop he had lost 12 minutes on the schedule and was lying third overall; with Edward Harrison in fifth place and several other Zephyrs still going well, the team prize at least looked a possibility.

After leaving Nairobi once more for the northern stage, Scott made up about 10 minutes to the control at Thika, where exactly 20 minutes was taken to replace the bent track rod at the Ford garage. This loss was made up over the next 50 miles and the Scott/Davies car pulled into the Kangondi control on time. The pace had now taken its toll, almost half the competitors were out, and Edward Harrison had moved up into fourth place after the Mercedes lying between himself and Denis Scott had retired

Jeff Uren on his way to 13th place overall in the 1959 Monte. Photo from the Jeff Uren collection.

with mechanical failure, although Mercedes were still holding first and second overall. With just 100 miles of the 3,200-mile route left to cover, the second-placed Mercedes went out with brake failure, so handing the team prize to the Zephyrs. Denis Scott and Edward Harrison had taken second and third place overall at their first attempt in the world's toughest rally, and had combined with the locally entered Zodiac of W.F.C. Young and L. Baillon to take the team prize for Ford.

In view of the Safari entry at Easter, the Competitions Department did not support the Tulip Rally later in April, although several Zephyrs were there, in some cases with works drivers. At the end of the 2,000-mile run, which had taken them as far south as Avignon before turning north towards the eventual finish in Noordwijk, the Zephyr of Peter Riley had taken third place overall behind a 3.4-litre Jaguar and a Triumph TR3, had won the 2,600cc class from Schock's Mercedes 220, and had combined with the Zephyrs of Cuth and John Harrison and of Gerry Burgess and Sam Croft-Pearson to give the Zephyr its second major team award in a month.

A full works Zephyr entry contested the Alpine Rally in June. The classifications had been changed for this year's event, with all cars over 2,000cc being in the same class, but with each capacity class divided into Touring (saloons), or Grand Touring (sports cars). Out of the 58 starters, only 27 survived to finish, and the Zephyrs continued their recent successes with the cars of Riley/Pitts, Harrison/Harrison and Harrison/Fleetwood coming in first, second and third in the over-2,000cc Touring class. This performance also gave the Zephyr its third consecutive team award in a major international event and added three more of the coveted Alpine Cups to the Zephyr's total. Anne Hall was particularly unlucky in this event, being deprived of the Ladies Cup almost right at the end when her Zephyr went out with mechanical trouble. None of the other ladies had survived as far as Anne Hall in this tough event, and the organizers presented her with a special trophy, for 'Courage, endurance and tenacity of purpose'. The severity of an event such as the Alpine can best be imaged

On Safari in 1959. Left to right, Edward Harrison, David Markham (co-driver), Bob Aston (Mintex), Competitions Manager Edgy Fabris and mechanic Jack Welch. Photo from the Edgy Fabris collection.

when it is realized that many of the passes were gravel-surfaced, often too narrow for overtaking, and some of the seemingly endless succession of hairpin bends were tighter than the turning circle of cars like the Zephyr. The technique used to get the Zephyrs round these particular hairpins was known as 'kicking it round'; first gear would be engaged on the approach and the tail end deliberately broken away with a wide throttle opening at the apex of the corner, then the car would be 'caught' by the driver as it

The camera catches Denis Scott's Zephyr as it dives into a huge pothole on the 1959 Safari whilst on its way to second place overall. Photo from the Denis Scott collection.

Anne Hall nears the top of the Stelvio on the 1959 Alpine Rally. Photo from the Anne Hall collection.

lined up for the exit. When talking about this method, recently, Gerry Burgess remembered an incident when, at one control point, all the Zephyrs were featuring a similar dent in one of the rear doors, after they had all collided with the same piece of rock jutting out from the rock face at the exit from one hairpin.

Saloon car racing was by now a popular part of the motor racing scene, and having been elevated to championship status the previous year it was also now being taken very seriously. To chart the Mk 2 Zephyr's success in this sphere of competition it is necessary to go back to 1957. Jeff Uren was successfully racing an Anglia that year, and he approached the Ford Competitions Department about the possibility of driving a works Zephyr in these events, a request which was turned down as there were already enough works drivers, and in any case, although competing in some of these races, the Competitions Department were concentrating far more on rallying. Although he didn't own a Zephyr at this stage, Jeff Uren nevertheless entered one for the saloon car race at Silverstone in September, and once the entry was accepted he bought a brand new example. The engine was immediately stripped down, and the individual components balanced before reassembly. A single $1\frac{3}{4}$in SU carburettor was adapted to operate as a downdraught instrument and replaced the standard Zenith. The Zephyr was run-in by driving from London to Exeter and back twice in one night the week before the race, and at the weekend, with just 700 miles on the clock, it took its place on the starting grid. In the practice session Jeff had lapped consistently slower than the works

Zephyrs, which were also competing that day, but in the race itself he comfortably led them home, being beaten only by the 3.4-litre Jaguars.

In 1958, Jeff Uren contested the BRSCC Saloon Car Championship as a private entrant, but was beaten on several occasions by Jack Sears in the BMC works-prepared Austin Westminster, which proved itself to be a faster car than the Uren Zephyr, although the pair did have some memorable dices that year, including one at Silverstone in May, when they lapped just feet apart for almost the entire distance and the Zephyr's superior handling qualities got it to the flag ahead of the faster car. Not far behind this particular struggle were three works Zephyrs with standard engines and automatic transmissions which were having a private dice of their own, eventually won by Denis Scott. The automatic Zephyrs had been entered at the suggestion of Borg-Warner, and in fact went remarkably well, their performance at Silverstone that day earning much favourable comment in the motoring press. Jack Sears went on to win the championship that year, with Jeff Uren runner-up.

Jeff was confident that he could win the championship with some works support, and he approached Ford once again. This time his approach paid off, and he was given facilities at Lincoln Cars and some financial support from the experimental engineering budget with which to prepare a racing Zephyr for the 1959 season. A Raymond Mays light-alloy cylinder head, with a compression ratio of 10:1, was used as a basis for the increased power output being sought. Three twin-choke Weber carburettors and twin fuel pumps, one AC mechanical and one SU electrical, supplied the mixture, whilst a Servais six-branch/twin-tailpipe exhaust system was chosen to complete the top-end arrangements. The engine was fully blueprinted, and an increased capacity sump incorporating cooling tubes ensured adequate lubrication at sustained racing speeds. These modifications resulted in 168bhp at 5,800rpm, and this was transmitted to the rear axle via the standard clutch and three-speed gearbox, and a heavy-duty propeller shaft with special high-torque universal joints. An alternative set of straight-cut higher ratio first and second gears was also available, as were several axle ratios from elsewhere in the Ford range, although the standard 3.9:1 was most used. The suspension was lowered all

BMC Competition Manager Marcus Chambers presents Jeff Uren with the Bonneville Trophy, which was donated by BMC for the BRSCC Saloon Car Championship. The scene here is at Brands Hatch after the final race in the 1959 series. Photo from the Jeff Uren collection.

Jeff Uren's Championship-winning Zephyr at the BRSCC Racing Car Show in January 1960. Photo from the Jeff Uren collection.

Jeff Uren demonstrates the cornering capabilities of another of his racing Zephyrs. Photo from the Jeff Uren collection.

Team Prize winners in the 1960 Safari. Left to right, Cuth Harrison and Peter Davies, Denis Scott, John Harrison and Vic Preston. Photo from the Denis Scott collection.

Zephyrs under preparation for the 1960 Alpine Rally. In an attempt to beat a new 'coefficient' handicap, which was heavily biased against the Zephyrs, these cars were fitted with Raymond Mays modifications which gave over 150bhp. Other modifications were the lower axle ratio from the Consul, which, in conjunction with a Laycock overdrive, gave a close-ratio six-speed gearbox, a high-ratio steering box (12.5:1), Girling disc front brakes and wider rear brake drums. To the right of the picture are Jeff Uren and Edward Harrison (both with cigarettes). Photo from the Norman Masters collection.

round by 1½in, with uprated Armstrong front struts 40% stiffer than standard embraced by 20% uprated springs. An extra leaf was incorporated in the rear springs, and Armstrong adjustable shock absorbers replaced the originals, with the rear axle additionally located by an anti-tramp bracket. The steering gear ratio was raised to 12.5:1, and the braking system was the servo-assisted Girling front disc/rear drum arrangement already used on the Alpine Rally Zephyrs. As there were no 13in racing tyres available, standard-size Michelin Xs running at 50-55psi on the standard Zephyr rims were used. Additional instrumentation, lightweight bucket seats, safety harness and perspex side and rear windows were the only bodywork/interior modifications. This proved to be a championship-winning combination, with six class wins and one second in the eight-race championship series, with class lap records at Aintree (2min 32.4sec), Goodwod (1min 53sec), Snetterton (2min 06.6sec) and Brands Hatch (1min 10.2sec). The final race in the series was at Brands Hatch on August 29, and Les Leston's Riley 1.5 could in fact still draw level, although only if Jeff Uren failed to finish and Leston won his class and set fastest class lap. However, it was Leston who fell back, while Jeff Uren completed an immaculate season to take the Bonneville Trophy which went with the BRSCC Saloon Car Championship and record yet another notable success in the competition career of the Mk 2 Zephyr.

Following this performance, Jeff Uren joined the Ford Competitions Department as Team Manager, with responsibility for looking after the driver's interests 'on the spot' during the events, a move which relieved some of the pressure from Edgy Fabris, who would now have more time to tackle the surprising amount of administration involed in organizing works participation in the major international competitions.

Additional fuel capacity was allowed on cars competing in the Modified or Grand Touring category. In order to enable the standard fuel tank, with its filler in the back panel behind the number-plate, to be filled from a jerrycan single-handedly, the Mk 2 rally cars carried a funnel with a hook which fitted over the lip of the back panel.

The last major event in 1959 was the RAC Rally, scheduled that year for November rather than its previous stagings in March. The new 105E Anglia was making its rallying debut in this event alongside the established Zephyrs, one of the latter being driven by racing ace Graham Hill, and another appearing in the hands of the regular pairing of Gerry Burgess and Sam Croft-Pearson. Winter arrived early in Scotland and the RAC Rally got well and truly caught up in the adverse conditions. From the control at Nairn the route led to Braemar, via the Tomintoul Pass, and now Gerry Burgess takes up the story: 'Arriving at Tomintoul we and the rest of the runners found the road hopelessly blocked by snowdrifts. Approaching there, we met many cars already going back the way we had come. It was fairly obvious that some of the cars had turned back even before reaching the block, and most we could see were turning west to come right round back to Braemar via the main road. I decided to turn east and whilst making as good time as possible Sam worked out our route through Dufftown, Ballater and into Braemar from the right direction. As soon as I began to get figures from him I saw it was possible to make the control

Anne Hall's Zephyr after an incredible high-speed loop on the Acropolis Rally in 1960. The Zephyr had run into a roadside quarry and climbed a vertical rock face before falling over backwards. Photos from the Norman Masters collection.

Mechanic Norman Masters (top) and electrician George Cheeseman prepare Gerry Burgess' Zephyr for the 1961 Monte.

before exclusion time and I got down to hurrying. In my memory, we flew; the Zephyr's wheels didn't touch the road again and we finally reached the control a few minutes late only, and a few minutes quicker than Levy, who was the fastest round the main road way. When I was told later that I was leading the rally it was hard to believe, but I was.' Gerry Burgess held on to his lead, despite a collision almost right at the end when the Zephyr hit a competing Aston Martin which was inexplicably coming the wrong way. Unfortunately, the German entrant Levy protested at the Zephyr's win, and asked for the Braemar stage to be deleted from the rally as many competitors had missed it out altogether. Sir Hartley Shawcross was appointed to hear the protest, and after listening to Levy and Ford Team Manager Jeff Uren he dismissed the German's argument and awarded Gerry Burgess his rightful victory.

After the successes of 1959 there were high hopes of a Zephyr victory at Monte Carlo in 1960, and at the end of the 2,000-mile road section the works-entered cars of Cuth Harrison, Edward Harrison and Denis Scott

Below left, triple SU carburettors on a works Zephyr rally car. The chain from the bulkhead to the radiator is for the radiator blind. Below, extra instrumentation in the 'Low Line' rally Zephyrs was carried in a centre console designed by Ken Armstrong.

were amongst the 10 unpenalized survivors, and had booked their place amongst those which would contest the final mountain circuit. However, in the mountains Edward Harrison went out when his Zephyr sustained a fractured brake pipe, Denis Scott left the road and Cuth Harrison surprisingly lost time. Meanwhile, the factory-entered Mercedes team, which, with typical German thoroughness, had spent the previous three months practising in the mountains above Monte Carlo, swept in to the finish for an impressive first, second and third overall. It was two privately entered Fords, the standard Zephyr of Mike Sutcliffe, who finished sixth overall, and the modified Zodiac of D. Handley, who won the over-2,000cc GT category, which kept the Ford flag flying at the end. After the disappointment at Monte Carlo, consolation came at Easter when the three works Zephyrs of Vic Preston/John Harrison, Cuth Harrison/P. Davies and Denis Scott/L. Baillon took the team award once again in what proved to be the toughest East African Safari yet, with only 25 cars out of the 84 starters completing the 3,200-mile course. Although Mercedes had again won the event outright, this was nevertheless a good rally for Ford, as in addition to the Zephyr's team prize the new Anglias had come first, second, third and fifth in the small car category.

The Harrison family had a keen interest in the 1960 Alpine Rally; with two consecutive Alpine Cups to his credit, Edward Harrison was in line to win one of the extremely rare Gold Cups, which were awarded for a third successive unpenalized run, and his father, Cuth, with two non-consecutive Cups, would win a Silver Cup if he, too, could finish unpenalized on the road this year. However, new 'coefficient' regulations, based on engine capacity and weight, had been introduced. According to Team Manager Jeff Uren, they were 'totally anomalous', being biased heavily against the Zephyr, and they made the 'impossible' schedule on this event just that as far as the Zephyr was concerned. Edward Harrison, particularly, made a terrific effort to beat the handicap, but just failed to do so and, although four Zephyrs finished, a brilliant run in this competition had been brought to an end. As if this were not enough, new regulations for the 1961 Monte Carlo Rally included a 'factor of comparison' which made a complete mockery of the event that year, and was seemingly introduced to ensure that no matter how well a car performed in the rally, if it were not a small-

Anne Hall on her way to a magnificent third place overall and the Ladies Cup on her first Safari. Photo from the Anne Hall collection.

For the fourth year running, three Mk 2 Zephyrs had the lowest aggregate penalties of any three-car combination on the Safari. Here the 1961 Team Prize winners line up. Anne Hall (sunglasses), here with her co-driver Lucille Cardwell, rates this as her best-ever achievement in a rallying career which brought her many awards at International level. In the centre is veteran motoring correspondent Tom Wisdom with his local co-driver and, on the left, Cuth Harrison and Peter Davies, who were sharing in the Team Prize for the second year running. Photo from the Anne Hall collection.

engined French model it would not figure high in the final placings. Nevertheless, the few big Fords which entered went well, with D. Handley's Zodiac, the lone works Zephyr of Gerry Burgess and Eric Jackson's Zephyr taking first, second and third, respectively, in the over-2,000cc GT category.

The East African Safari was not affected by such things as 'coefficients' and 'factors of comparison'; it was a case of survival of the fittest in a straight fight in which the terrain itself was the 'handicap' which ensured that many would not finish. Although a team of Anglias was entered once again for the 1961 event, Anne Hall, who already had three Ladies' Cups to her credit in the new Anglia, was back at the wheel of a Zephyr for this, her first Safari. Soon after arriving in Nairobi, Anne Hall, with her local co-driver Lucille Cardwell, was out on a reconnaissance of the southern loop when the Zephyr hit a deep drift at speed, ripping off both front wheels and bringing the car sliding to a stop on the remains of its front suspension. The hazard had quite probably been deepened by the local natives, who liked to see cars crash, and would then offer 'protection' of the vehicle on the payment of a few shillings whilst the driver went for help. The protection was necessary, otherwise such things as tyres would be cut up for sandals and many other items from the car would be put to all sorts of uses which the manufacturer had never dreamed of. This incident highlighted the value of a local co-driver for the British entries, as Lucille Cardwell could speak some Swahili and soon 'organized' things with the natives, including the loan of a bicycle. Seven miles of pedalling brought Anne Hall and Lucille Cardwell to a large village, from where they were able to hitch a ride to a town and eventually arrange for transport back to Nairobi and the recovery of the Zephyr. Now without a recce car of her own, Anne Hall went over the northern loop with Jeff Uren in one of the Anglias. On the rally itself, the southern loop didn't give Anne Hall any trouble; she did lose some time, but so did everybody else, and many were already dropping out. As the survivors returned towards Nairobi for an overnight rest before tackling the northern loop, the familiar Safari Mercedes-versus-Zephyr picture was emerging once again.

At Nairobi, two past winners, local drivers Manussis and Fritschy, were lying first and second in their Mercedes 220SEs, being 22.86 and 26.58

The winning Zephyrs joined by the successful Anglia team which dominated the small-car category for the second year in succession. In front of car number 38 is Jeff Uren (in dark shirt). Photo from the Jeff Uren collection.

minutes late, with Anne Hall in third place having lost 35.52 minutes. With the Zephyrs of Cuth Harrison and veteran motoring correspondent Tommy Wisdom being close behind with 41.84 and 44.20 minutes lost, the big Fords were once again well placed for the team award, although anything could still happen on the 1,300-mile northern loop. After an overnight rest, the rally was underway again. The tightest section of all on the northern loop was a special stage from Rondo to Timboroa, 55.2 miles of rough mountainous road which would take the competitors through several small villages with 30mph speed limits, enforced by a secret check in one of the villages. Any time over an hour taken on this section would mean more penalty points, and recalling this stage recently Anne Hall remembered: 'We put crash hats on and set off; it was a case of driving like lunatics.' On one of the very few bits of decent 'road' on this section, sharp stones had been strewn across the road by the local natives and Anne Hall's Zephyr sustained a puncture. The car was quickly surrounded by grinning natives who simply lifted the offending corner of the car clear of the ground while the crew did a record-breaking wheel change without the need of a jack. The Zephyr covered this stage almost exactly on time. Anne Hall didn't give either Manussis or Fritschy a chance to ease up at all, and was herself being pushed hard by a Humber Super Snipe close behind in fourth place, which was also going extremely well. Towards the end of the rally, coming over the brow of a hill at over 70mph, Anne Hall was faced with a right-angle turn over a narrow bridge which was much nearer the brow than she remembered from the recce. 'Luckily, I had practised handbrake turns', she remembers, and the Zephyr arrived sideways at the bridge scattering a group of reporters gathered there. The story goes that when the Zephyr appeared at speed over the hill, a chorus of 'She'll never make it' went up from the assembled scribes. One of them, however, who had seen her in action in Europe, quickly replied 'I'll bet everybody here

£10 that she does'. She did, but whether the opportunist gambler ever collected his winnings is not recorded. She kept going hard to finish third overall and to win the Ladies' Cup in her first Safari, coming in 13 minutes and 7 minutes behind the first and second place Mercedes. Tom Wisdom and Cuth Harrison came in sixth and seventh overall, and this, coupled with Anne Hall's performance, gave the Mk 2 Zephyr yet another team prize in the world's toughest rally.

After this, the Competitions Department turned its attention more to the Anglia for the immediate future, as the Mk 2 Zephyr was, of course, well into its last year of production. The big Fords had completed a remarkable run in international rallying, having contested, as far as can be ascertained, 48 events, many as private entrants only, which had resulted in four outright wins, six team prizes, six Alpine Cups, four Ladies' Cups and 16 class wins, in addition to their disputed win and team award in the 1958 Safari. This was a record second to none throughout the period, and the Zephyrs particularly had proved themselves to be at least the equal in ruggedness and durability of cars like the Mercedes 220, which was selling at three times the price, and was considered by many to be the yardstick by which medium/large six-cylinder saloons were judged. That all this was achieved on what amounted to a shoestring budget does the greatest credit not only to the cars themselves, but also to Edgy Fabris; Jeff Uren and the team of mechanics at Lincoln Cars and, of course, to the skilled touch of the drivers to whom the Zephyr had endeared itself. Together, they had laid the firm foundations on which were to be built the great competition successes of the Cortina and Escort in later years.

CHAPTER 5

Obsolescence and renaissance

From bargain basement to collectible classic

The declining years

By virtue of their sturdy and straightforward design and construction, backed up by a spare parts and servicing organization that was widely acknowledged to be second to none, the products of the Ford Motor Company had built up an enviable reputation for reliability and long life. No matter where his travels might take him, the owner of an old Ford was secure in the knowledge that if anything at all should go amiss with the car, there was likely to be a garage near at hand displaying that familiar dark blue oval sign that indicated the premises of a Ford Agent. With this sort of backing there was always a demand for secondhand Fords, and the recently obsolete model could be a particularly attractive proposition for the used car buyer.

The immediate sales success of the big Mk 2 Fords inevitably produced a steady flow of Mk 1 models amongst the part exchanges that were taken in, and these were soon appearing on the dealers' forecourts at very tempting prices. Typical of what was on offer during the summer of 1956 were: '1954, June, Ford Consul saloon, black, red leather, heater, overriders, 22,000 miles only, £545 secures', at W. J. Brown Ltd, of Finchley Road; '1954 Zephyr saloon, genuine 14,000 miles only, £550', at Ripco Ltd, Mayfair; and '1954 Zodiac, radio, one owner, absolutely as new, £595', at Scott Cars, of Hampstead. These particular advertisements, accompanied by many more like them, appeared in the classified columns of *The Autocar,* and similar offerings could be found in the 'Cars for Sale' columns of newspapers throughout the country.

The announcement from Dagenham on July 20, 1956, of the introduction of 'The Ford "A1" Used Vehicle Plan', with its comprehensive guarantee, to be operated by Ford Main Agencies, added considerably to the attractiveness of the recently obsolete Mk 1. At a mere 2 or 3 years old, a Mk 1 Ford would usually still be in excellent all-round condition and at the sort of prices quoted could represent a considerable bargain, offering a degree of comfort, refinement and performance that could not possibly be matched by a brand new 'economy' model. In fact, at this stage, good used Mk 1s were rapidly being snapped up, and this resulted in the prices hardening over the next 2 years or so; as late as 1959 a clean 1954 Zephyr could easily fetch £450, and a convertible like the 'Zephyr convertible '55, power hood, 2-tone blue/white, low mileage and in really outstanding condition; £525', at B & M Garages, Paddington, would be considered excellent value.

Many Mk 1s, however, were still in the hands of their obviously well-satisfied original owners, some of whom took the trouble to write to the Ford Motor Company to record their appreciation of the excellent service and reliability that their cars had given. So impressed with his 1955 Zephyr was G. E. Bishop, AMIEE, of London N12, that he asked: 'Why did you stop manufacture of such an excellent model?' After having a reconditioned engine fitted into his 1953 Zephyr, V. H. Allan, of Sunderland, wrote: 'You may be interested to know that the original engine ran faultlessly for 109,520 miles. The petrol consumption was 4,385 gallons, mostly of the cheapest grade, giving an average of 24.98mpg on predominantly short-distance runs. It was still running perfectly when changed, but the oil consumption had gradually increased from nil at 79,000 miles to about 160mpp at the end.' Two other high-mileage Mk 1 models were being featured in oil company advertisements at about this time. A 1955 Consul, owned by J. E. Piper, of Ipswich, had covered 105,806 miles without the cylinder head ever being removed, and had always been lubricated with BP Visco Static. The highly tuned 1953 Zephyr that was successfully raced and rallied by Leeds businessman Jack Reiss had by now amassed 143,000 miles of use on road and track. Castrol had always been used in the sump, and the only work ever carried out on the engine had been the replacement of the piston rings at 90,000 miles.

Apart from some of those which had suffered serious accident damage and had thus been considered beyond economic repair, very few Mk 1 cars had been scrapped by 1960, and with replacement parts, including body panels, still readily available at typically low Ford prices, many early examples of the model still had a considerable lease of life ahead of them and were continuing to give good service after, in some cases, having passed through the hands of several owners. Indeed, the popularity of the Mk 1 cars was such that in May 1960 the Ford Motor Company published an entirely new handbook, covering all three models, in response to many requests from owners of secondhand examples from which the original handbook was missing.

As the 1960s got under way, however, several factors were soon to combine and, between them, help to bring about a rapid decline in the number of early postwar cars on the road. Section 65 of the Road Traffic

The author's 1955 Zephyr Zodiac at a collectors' car display in North Yorkshire in 1978. Photo by the author.

Act, 1960, required any private car, or light commercial vehicle weighing under 30cwt unladen, that was 10 years old or more to be submitted to an authorized testing station, where its 'braking systems, steering gear, lighting equipment and reflectors' would be examined. The testing station would be an already established garage where an authorized member of the staff would carry out the examination on behalf of the Ministry of Transport. If a vehicle failed to meet 'certain statutory requirements' it was still permitted to be driven to a place where the necessary repairs could be carried out, and it could then be submitted for examination again, and a test certificate would be issued. In practice, many owners of a vehicle which had failed its test simply tried their luck elsewhere, and such were the discrepancies existing between testing stations in the early days that this method would often result in success.

It had long been the policy of the Highways Departments of the Local Authorities to combat the dangers of icy conditions in the winter months by liberally applying sand and grit to the affected roads. In extreme cases, particularly in some rural areas, salt would also be used, its ice-melting qualities being particularly effective against severe conditions, but in towns and cities its use so far had been limited and relatively few cars had been exposed for long to its disastrous corrosive effects. In the big freeze-up of the 1962-63 winter the use of salt appeared to increase dramatically without any thought being given to the deterioration that it would cause to road-going vehicles, and from this time on its use became ever more widespread throughout the winter months.

The big Fords of the 1950s were by no means immune from the effects of these measures, and by now, early examples of the Mk 1 models were being submitted for the MoT test. Failures were not uncommon, particularly with regard to the steering gear, where the track control arm ball-joints seemed to be a favourite target for the testers' dissatisfaction. From the owners' point of view, with the parts available over the counter at Ford's low prices, the failure was more of a nuisance than a disaster, and the MoT test would not seriously deplete the ranks of the Mk 1s just yet.

With the introduction in April 1962 of the Mk 3 Zephyr and Zodiac range, the big Mk 2 Fords were now relegated to obsolescence alongside the earlier cars. There were some excellent deals to be had amongst the few remaining unsold Mk 2s, and within a very short time most of these had taken to the road. By 1965, £450 would buy a virtually 'mint' late-model 'Low Line' Zodiac, and for those motorists who were seeking roominess with economy there were many excellent low-mileage examples of the Mk 2 Consul available in the £350-£400 bracket. At the other end of the scale, however, the Mk 1 cars were beginning to disappear from the scene. Corrosion was now a serious problem in many of the early examples, and the scope of the MoT test was being extended to embrace younger cars. At a time of increasing affluence, when an increase in wages meant a real improvement in a person's living standards, the values of old cars fell dramatically, and an MoT failure which called for a steering gear overhaul and perhaps a pair of tyres could well mean the sudden demise of the car in question, its market value being such that the cost of the necessary repairs could not be justified.

The majority of Mk 1 cars were now very definitely in the 'banger' category, and a typical advertisement in the 'Cars For Sale' column of the local newspaper would read something like: '1954 Zephyr, 8 months MoT, good runner, £75'. Anyone answering this sort of advertisement was likely to come across a Zephyr that looked reasonably sound, but close

A well-kept Mk 2 Zodiac photographed at the same display as the author's Mk 1. Photo by the author.

inspection would soon reveal evidence of advanced corrosion in many areas. The front wings would be suffering quite badly along the rear edge adjacent to the door, and again at the top rear corner where attempts may well have been made to rectify the situation with body filler. Feeling along the bottoms of the sills would reveal that the inner and outer sections were no longer joined together, and the rear wings would more than likely have come adrift at their front lower mounting point. On getting into the car the prospective purchaser could well be confronted by a mileometer reading of 80,000 to 90,000, and the condition of the driver's side of the front bench seat would confirm this with its broken stitching having allowed the separate upholstery panels to come apart in places and create a particularly tatty impression. The carpets, too, would be well worn in the front compartment, but looking to the rear would reveal a better picture, where both carpets and seats had been subjected to rather less wear and tear. The cloth roof lining always seemed to discolour, taking on a darker and often patchier appearance than when new, and this also detracted from the interior scene. However, the window winding mechanism, door locks and handles, and the facia-mounted switchgear were all of good quality, and were still likely to be working well with a nice firm feel about them.

Starting the engine would reveal some tappet noise, and there would be at least some traces of blue smoke in the exhaust, even in a well serviced unit. On the road the engine would pull well and pick up rapidly in response to a suddenly opened throttle, one of the most endearing features of the old straight-six being its lack of temperament and its ability to deliver top performance after a considerable mileage had been covered. The column gear-change would still be accurate, but the synchromesh between top and second gear would probably have weakened by now, slowing the downward changes. Clutch judder would almost certainly be evident when moving off in reverse.

Being in possession of an MoT certificate, or 'ticket', as that document had now become widely known, should ensure that the running gear was in a reasonably safe condition. The shock absorbers seemed to be of a particularly sturdy construction and specification, with the result that the 'banger' Zephyr would usually ride extremely well, although some weakening of the springs might be evident in rather more lean during fast cornering than would be encountered with a younger car, and more pronounced 'nose diving' under hard braking. A bid of £60 would probably be sufficient to see the old Zephyr undergo a change of ownership once again, quite probably for the last time.

The fact that many Mk 1 models were now languishing in breakers' yards meant that an additional source of cheap spare parts was available for those motorists who were perfectly happy to be running an old car, and this situation helped to prolong the life of some of the surviving cars. But with the Mk 2 models beginning to drop towards the bottom end of the used car market, the end was near for many Mk 1s. In the booming economic climate of the late 1960s few people were interested in running old cars, and even very well kept examples would fetch little more than a 'banger' when offered for sale. With tempting hire purchase facilities available on newer vehicles, many dealers would inflate the prices of their stock of recent cars and then offer an inflated part-exchange allowance on the older car and thus allow its owner to move up to a much newer model without having to find a cash deposit. The old, part-exchanged car would then as likely as not disappear round the back of the dealer's premises to await the next scrap merchant's visit. As this practice became widespread, many excellent designs of the previous decade were virtually wiped out, including the Consuls, Zephyrs and Zodiacs. By the early 1970s the Mk 2 models were becoming a rare sight. To many of their former owners, the big Fords of the 1950s were nothing more than a snapshot in the family photo album, which in years to come would serve to bring back some very pleasant motoring memories.

Classic status

Early in 1977, Dave Barry, of Burgess Hill in Sussex, placed an advertisement in the 'Collector's Cars' column in *Exchange & Mart* enquiring whether anyone was interested in forming a Mk 1 Consul, Zephyr and Zodiac owners' club. Several favourable replies were received, and the nucleus of the club was quickly established. A meeting was arranged, and on Sunday, April 3, 1977, some 15 Mk 1s assembled in a leafy Buckinghamshire orchard at the home of Ford enthusiast Dave

Mk 2 model pictured at a Northern area Mk 1/2 Owners' Club meeting in 1980. Photo by the author.

A line-up of Mk 1 models at a Mk 1 Owners' Club meeting in Buckingham. Photo by the author.

Cropper. Applications for membership were steadily increasing, and an ambitious London-to-Brighton run was organized by brothers Dave and Dick Barry, under whose guidance the Mk 1 club was now gathering strength. On the sunny Sunday morning of May 15, 1977, over 30 gleaming Mk 1s travelled from points all over the south of England and from as far north as Yorkshire to assemble on the Carriage Road in London's Hyde Park, ready for the run to Brighton. The run was a big success, good publicity was obtained and many new friendships were formed. A handful of keen Mk 2 members joined in the Brighton run, and it was one of their number, Dave Debenham, of London, who decided the time was ripe to launch a Mk 2 club.

Membership of both clubs continued to rise steadily, and in 1978 Mk 1 club founder member Dave Barry travelled to Australia where he made contact with the already established Consul Zephyr Zodiac Club of Melbourne, and the Consul Zephyr Zodiac Club of Queensland, before founding a club in Sydney from which he then set about building the Consul/Zephyr/Zodiac scene into an associated international organization in which, although the clubs each remain autonomous, they nevertheless keep in contact with each other.

Amongst the owners' club membership there are many outstanding examples of both Mk 1 and Mk 2 models, although unfortunately none of the works rally cars seem to have survived: it seems it was Ford policy at the time to break up these cars once the Competitions Department had no further use for them. However, thanks to the efforts of David Burgess-Wise of Ford Motor Company Public Affairs, a Mk 1 Zephyr owned by the Ford Motor Company is now resplendent in the colour scheme of Maurice Gatsonides' 1953 Monte Carlo Rally-winning car, and is complete with the anti-glare shield over the fog lamps, rally plates and the authentic registration number VHK 194, after recently being re-registered. Whilst the 'new' VHK 194 represents the Mk 1 Zephyr's past competition glory, Mk 2 Zephyrs and Zodiacs can be seen in action again in the popular 'Pre-1957' saloon car races being staged at many circuits. These events are organized by the Classic Saloon Car Club, and are restricted to saloon models which were introduced into production prior to December 31, 1956, thus just letting in the Mk 2 Zephyr/Zodiac which, as can no doubt be imagined, has proved to be a formidable contender. In addition to modifications directly affecting driver safety, such as FIA-approved roll-cages, foam-filled alloy

Some 20 years after the Mk 2 Zephyr's competition heyday, Eric Walker speeds his Zephyr round the Silverstone circuit in a Classic Saloon Championship event. Photo from Eric Walker.

fuel tanks and safety harnesses, modifications to improve the performance are also permitted, and the specification of a Mk 2 Zephyr which has been successfully raced in these events by Eric Walker makes an interesting comparison with Jeff Uren's championship-winning car of 1959.

As the cylinder head material has to be original, the Raymond Mays light-alloy pattern is excluded, and the Eric Walker car features a reworked standard head with opened-out ports and larger valves. The block is bored out to the maximum 60 thou, which just allows Mahle pistons to be used, these, along with the connecting rods, crankshaft and flywheel, being balanced. Triple Reece Fish, or alternatively triple SU carburettors are used, along with a six-branch exhaust system. The gearbox is based on a Mk 3 four-speed, with a combination of Granada and Zodiac internals and a Transit van floor gear-change, whilst the suspension has uprated Spax front struts and Spax adjustable rear shock absorbers. The car runs on 7in-wide rims with 185 × 13 70-series Goodyear G800S tyres. In certain respects, particularly with the gearbox and wide wheels, this specification is further removed from standard than was the case in saloon car racing in the 1950s, although the bodywork of these Pre-1957 saloons has to be strictly standard and include the original-style bumpers. Mk 2 models such as the one described have enjoyed considerable success, and a Zodiac driven by Terry Heley won the Pre-1957 Saloon Car Championship in 1980.

In February 1981, the Zephyr Zodiac Club of Christchurch, New Zealand, hosted a Consul Zephyr Zodiac Convention to mark the 25th anniversary of the introduction of the Mk 2 models, which was attended by many members of the Australian clubs. In Britain, Mk 1 club membership totals approximately 250, the affairs of the club now being handled by Dick Barry, who also looks after the club spares, the stock of which sometimes includes surplus items shipped over from Australia. The Mk 2 club has grown to some 600 members and their affairs are still kept in order almost single-handedly by the energetic Dave Debenham. Both clubs hold regular meetings and each has exhibited at the Classic Car Show in recent years. Indeed, it is largely through the efforts of the hard-working organizers of these two clubs that the Mk 1 and Mk 2 Consul, Zephyr and Zodiac have at last gained the 'Classic' status which they deserve.

CHAPTER 6

Owning a Big Fifties Ford

Performance, purchase and maintenance

Performance in the 1980s

To those enthusiasts who are fortunate enough to be able to possess more than one car, living with a car from the early postwar era should not present any serious problems. If the car is only to be used and displayed as a beautifully restored or preserved showpiece at summer shows or owners' club meetings, then it matters very little if there is a lack pf performance by present day standards. If, however, the old car has also to serve as everyday transport under all conditions, then the performance and general road behaviour will play an important part in the enjoyment of ownership.

The least powerful of the big 1950s Fords, the Mk 1 Consul, can exceed by about 5mph the maximum speed of 70mph that is permitted by law in Britain today, and is capable of a rate of acceleration that will allow it to maintain good progress under all conditions; if handled intelligently, it should in no way be the cause of unreasonable hold-ups and the consequent embarrassment to its driver. In terms of sheer acceleration, however, it is now well down in the performance league; taking 25.1 seconds to go from 20 to 60mph (*The Autocar,* July 22, 1955) using second and top gear compares rather unfavourably with a Fiesta 950 at 18.7 seconds (*Autocar,* February 5, 1977) using second and third gear, or 23.1 seconds using third gear only. Using just top gear, the flexible nature of the Consul's engine makes it considerably livelier than the Fiesta over the same 20 to 60mph range, requiring only 29.9 seconds against the Fiesta's 39.7. The Mk 2 Consul does a little better all round than its predecessor, and its times over the same speed range of 21.7 seconds (*The Autocar,* November 15, 1957) in second and top gear, and 26.3 seconds using top gear only, indicate its capability of putting up good cross-country averages without requiring hard work from its driver. Indeed, by the four-cylinder standards of today, its flexibility is excellent, and this helps make it a most pleasurable and easy car to drive in whatever the prevailing traffic conditions.

An engine capacity 50% greater to cope with a weight increase of less than 10% ensures a very lively performance from the six-cylinder models. Within the limits of its by now rather modest maximum of 80mph or so, the Mk 1 Zephyr is still a brisk performer, and that ability to climb a 1 in 20 gradient at a full 70mph in the same top gear that requires only 8 seconds to go from 20 to 40mph is quite sufficient to guarantee a high average journey speed if the driver so desires. Overtaking slower-moving vehicles presents no problems, and the times taken in top gear to go from 30 to 50mph and 40 to 60mph of 9.1 seconds and 11.3 seconds, respectively, (*Motor*, R/T No 16/55) make an interesting comparison with the figures of 10.4 and 11.1 seconds, respectively, recorded by a Granada 2.3-litre (*Motor*, R/T No 45/81). As the Granada model in question is also the nearest equivalent to the Mk 2 Zephyr in the current Ford range, comparing the two now will put the performance of the most powerful of the 1950s Fords into today's perspective. The Mk 2 Zephyr's time for the standing-start quarter-mile of 20.3 seconds (*Motor*, R/T No 19/58) compares unfavourably with the Granada's 18.3 seconds, but when it comes to top gear passing times the picture changes considerably. The Zephyr's times from 20 to 40, 30 to 50, 40 to 60 and 50 to 70mph of 7.8, 8.3, 9.3 and 11.3 seconds, respectively, indicate a clear margin of superiority over the Granada's comparable figures of 10.4, 10.4, 11.1 and 11.8 seconds. In fairness to the Granada, it has to be said that above about 75mph it carries on much more strongly than the Mk 2 Zephyr, and its ultimate maximum speed of 107mph is over 15mph better than the earlier car. The figures that have been quoted are in no way intended to belittle the Granada's performance, but rather to point out that at the middle-range speeds that the majority of drivers use for much of the time, the Zephyrs and Zodiacs of the 1950s can still give an excellent account of themselves.

Purchase and restoration

The enthusiast who wants to acquire a big Ford from the 1950s has a number of choices to make. In addition to deciding which version, four- or six-cylinder, and which model would be most suitable, it pays to define at the outset just what is the ultimate goal and what are the minimum standards acceptable at the time of purchase. If, for example, a 'concours' restoration is the eventual aim, it will be necessary to choose far more carefully than if a somewhat less ambitious end result will be considered satisfactory.

Once these decisions have been reached, the search can begin. The scarcity of these models, particularly the Mk 1 range, means that the potential buyer will be very lucky indeed to pick up the local evening paper and find just what is wanted in the neighbourhood. A copy of *Exchange & Mart* might well reveal something, as could the classified columns of *Practical Classics, Thoroughbred & Classic Cars* or *Classic and Sportscar*. A letter to the respective owners' club secretary, enclosing a stamped, addressed envelope, will bring details of any that are for sale amongst the club membership. When answering an advertisement by phone it is advisable to have a written list of questions ready, for it is surprising just how many important points can be completely forgotten in the initial excitement of finding a potential purchase. Before arranging to undertake a long journey in order to inspect the car, it is as well to make sure that it at least meets the basic requirements already decided upon. If it transpires that the vehicle is a non-runner it is usually better to leave it alone: there is no really satisfactory way to make a reliable assessment of the major mechanical elements and the running gear if the car is not driveable. Unfortunately, because of the recent upsurge in interest in the cars of the early postwar period, some people are advertising 'collectors items' which are in fact derelict cars that for some reason they simply never got around to having towed away. So that initial phone call should be used to establish just what category the car really falls into. If everything sounds promising at this stage, then make the necessary appointment, but still be prepared for a wasted journey; the mental picture that has been built up through studying the advertisement and talking to the vendor may well differ considerably from reality.

The basic structure of both models is extremely strong and, generally, long-lasting, so that it is possible to be confronted by an example that has certainly not been well looked after and is looking distinctly poor around the wings, door bottoms and so on, and yet on close inspection reveals a really sound and solid structure. The chassis members, floor, inner wings, inner sills, door pillars and suspension mountings must all be in a perfectly satisfactory condition and require no more attention than a steam clean and the application of a rust preventative. Bearing this in mind, if the car being inspected has suffered serious deterioration in these quarters, then think very carefully before going ahead with a purchase. Repairs may prove both difficult and expensive, particularly if the buyer has not the facilities or the skill to carry these out and will therefore have to pay for professional work. It may well be advisable to look elsewhere for a better example in that case. As far as the outer panels are concerned, proceed on the assumption that new replacements will almost certainly not be available, and assess carefully the possibility of repairing the existing ones. Secondhand wings and doors in good or easily repairable condition do crop up from time to time, and here again these will be located through advertisements in the appropriate magazines, and of course the owners' club newsletters. Scrapyards may still sometimes yield just what is required, although examples of these models with recoverable panels still languishing in these establishments are now very rare. If the car was originally undersealed, this treatment will have covered the bolts and their attachment points, making the removal of the bolt-on wings far more difficult than would otherwise be the case. The underseal will almost certainly have hardened with age and consequently cracked and peeled in places; check very carefully where this has happened, as damp that is held behind loose patches of underseal causes unseen corrosion at an alarming rate.

If a 'concours' example is the eventual aim, the condition of the interior, and its originality, are most important. Bringing a shabby interior up to a high standard is, of course, possible. Trim materials, upholstery paints and carefully colour matched leather renovating kits can all be obtained from the specialist suppliers who deal in this aspect of car restoration, and there are upholsterers and coachtrimmers who can make and fit excellent roof linings and carpets. But, apart from the cost, the materials used will almost certainly not match the originals exactly, and details like, for instance, the texture of the roof lining cloth or the grained pattern on a PVC panel will vary. Therefore, if absolute authenticity is required, a car with a totally original interior should always be sought after.

If the car is in running order, and taxed and tested, then a trial run should soon reveal any serious mechanical weaknesses. The steering should be both light, particularly in the case of the Consul, and precise. In recent years this type of steering gear has come in for regular criticism in road test reports, particularly of Japanese products, for a general lack of precision and feel. The Burman steering box used on the 1950s Fords is similar to that used by Jaguar throughout the same period: it is fully adjustable for wear, which makes it extremely long-lasting as a rule. If the various steering joints, including the idler arm assembly, which is also adjustable for wear on these models, are in good condition, then a pleasantly light and accurate degree of control should be evident. The springs and shock absorbers were also hard-wearing components, particularly by modern standards, and a good ride under all conditions will often still be a feature of a very high mileage example.

Even from an engine in good condition there is likely to be some tappet noise, as the specified 14-thou clearance at the rockers does not really lend itself to silent operation. If the engine breathes heavily out of either the crankcase breather tube, mounted low down on the offside rear of the block, or the breather cap over the oil filler tube on the rocker cover, first check that the other breather has not become blocked. If one of these breathers does become choked with sludge it will cause excessive fuming at its opposite number. If, however, the breathers are obviously clear and heavy fuming is evident, then suspect serious bore, piston, or piston ring wear. Blue smoke in the exhaust does not necessarily imply excessive wear on these engines. If the diaphragm on the vacuum side of the combined vacuum/fuel pump has split, oil can be sucked out of the sump and fed into the inlet manifold; drawn into the cylinders, it naturally causes blue smoke in the exhaust. In extreme cases oiling-up of the plugs can occur because of this fault. An exchange pump or a vacuum pump overhaul kit will cure the problem. Oil leaking past the valve stem oil seals is not uncommon on high-mileage examples of this engine, and this can also lead to blue smoke in the exhaust, particularly when restarting a hot engine. If the engine checks out all right and is obviously in a sound and serviceable condition, then irrespective of the mileage covered it can still be expected to have a considerable lease of life ahead of it and, subject to the correct setting of plugs, points and carburettor, the performance should closely match that of a 'just run-in' example.

Clutch take-up should be smooth in first gear for both gentle and rapid starts, but when reversing very slowly, particularly when the clutch is being slipped, some judder will almost certainly set in. If the gearbox is in good condition it will operate quietly and smoothly and have powerful synchromesh. On very high-mileage cars, particularly the six-cylinder models, the synchromesh will probably have weakened, necessitating a slight pause on the downward change from top to second gear. Occasionally one may also come across a Zephyr in which second gear will jump out of engagement on the overrun. This is not usually indicative of any imminent disaster, nor is it often a serious inconvenience, as, thanks to the close proximity of the gearlever to the steering wheel, it is a simple matter to hold the lever in the second gear position. On all models excepting those Mk 2s produced between October 1957 and February 1959, which featured the concentric column linkage, the gear-change should be extremely light and positive, and when necessary, very quick. In sub-zero temperatures this gear-change can be quite stiff during the first half-mile or so after a cold start, but the normal lightness returns quickly as the car warms through. The concentric gear-change is rather less satisfactory, but although not as quick and accurate as the other arrangement, it should still not give rise to any problems as far as the driver is concerned. On the Mk 1 cars, particularly the Zephyrs and Zodiacs, there is often a characteristic rear axle whine. This usually only becomes evident when the car is fully warmed up and cruising on a light throttle, and is not in any case particularly obtrusive. On the Mk 2 models the final drive unit should be almost inaudible, and if not, considerable wear may have taken place.

Thanks to the generally good availability of mechanical components and service items, a serious mechanical defect should be relatively easy to rectify if the funds to cover this are available, and so if the car under inspection is suffering from some mechanical ailment it certainly need not be dismissed purely on these grounds, particularly if the bodywork and interior are in good condition.

Spares, servicing and maintenance

The Ford Agencies, motor factors and motorists' accessory stores cannot be expected to stock spares items for models that have been obsolete for 20 years or more. Whilst it can sometimes still prove worthwhile asking at some of these establishments to be permitted to search through their old stock, nearly all the parts that will be needed will have to come from either the specialists dealing in the make and model concerned, or the appropriate owners' club. In order to reduce the possibility of delays at routine servicing time, and if or when an MoT failure or a minor breakdown occurs, the owner should build up a stock of the expendable spares items that will almost certainly need replacing from time to time. Such a stock should consist of at least one full set of radiator and heater hoses, plus of course a fan belt, a set of engine gaskets, including a decoke set, and a fuel/vacuum pump repair kit and carburettor float and needle valve assembly. A complete set of spark plugs, contact points, condenser, rotor arm and distributor cap are all useful standbys in the event of a sudden ignition system failure. In view of the MOT test it is wise to keep spare steering joints in hand, and of course a set of relined brake shoes and brake cylinder overhaul kits. A complete clutch unit, including a release bearing and the hydraulic cylinder seals, will almost certainly be needed at some time, as inevitably will a new exhaust system. Front and rear hub seals, and a final drive pinion oil seal are also worth having ready, as can be a water pump, dynamo, starter motor and a pair of universal joints. At first glance this may seem a considerable list, but in fact it will not involve a very substantial outlay, and when viewed in the light of constantly rising prices it can actually represent a real economy. Its main purpose is to keep the car on the road by eliminating as far as possible the delays that can occur whilst waiting for a replacement part via mail order.

A servicing schedule based on the manufacturer's original recommendations should be carried out at regular intervals. Providing that a good quality molybdenum disulphide or lithium-based grease is used, the originally specified 1,000-mile intervals between attention with a grease gun can be increased, particularly if the car is only being used in good weather conditions, when a mileage interval of 2,500 should be quite safe. If, however, the car is used throughout the winter months and with plenty of wet weather running, the 1,000-mile intervals should not be exceeded. The 5,000-mile interval between engine oil changes that is recommended in the car's handbook should be regarded as a maximum, and if a low annual mileage consisting of predominantly short runs is being covered, then an oil change every 2,500 miles, with the filter element changed after 5,000 miles, is to be recommended in order to keep engine wear to a minimum. The Ford oil filter element (EOTA 6731-A) can be replaced by Unipart (GFE 124), Wipac (BE 411), Wix (77P), AC (FF 51), Fram (CH 885 PL), Purolator (MF 35 A) or Tecalemit (FP 3332) alternatives. Although it has an indefinite shelf life, engine oil deteriorates rapidly after becoming contaminated by combustion deposits, and therefore should not be kept in the sump for periods of much more than six months. Bearing this in mind, the enthusiast who is only covering a very small mileage may find it advantageous to work out a servicing schedule on a six-monthly basis, regardless of distance covered. Modern transmission oils retain their specified lubricating qualities almost indefinitely, and as they are not subject to the contamination that causes the rapid deterioration in the engine oil, periodic changes are now generally considered unnecessary. Following a gearbox or axle overhaul, however, it would be advisable to change the oil after a 350 to 500-mile running-in period.

If the sparking plugs, contact points, carburettor and air cleaner are cleaned and checked at 5,000-mile intervals, the performance should remain consistent, with no appreciable falling off between servicing. The original spark plugs, Champion N8, are now becoming rare, but Champion UN12Y, Autolite/Motorcraft AG5, or AG52 are all suitable replacements. The slightly cooler-running Champion N5 was preferred by some drivers, particularly if the car was habitually driven hard, and Champion N9Y, Autolite/Motorcraft AG4 or AG42, and Esso 8629 are similar alternatives.

Cars that are not actually laid up, but are nevertheless used infrequently, with periods of inactivity sometimes lasting weeks rather than days, can suffer serious deterioration in certain respects. The braking systems on the big Fords seem particularly prone to seizing up during even relatively short spells out of use, and this can sometimes lead to considerable expense as well as inconvenience if, for instance, a seized caliper on one of the disc-braked cars cannot be freed off and has therefore to be replaced. Problems such as this can often be avoided by spending just a few minutes with the car if it has been idle for several days and is unlikely to be used in the immediate future. Simply operating the brake and clutch pedals, and putting the handbrake on and off a few times will help to keep the linkages and hydraulic cylinders free and in good working order. After a week or more without being run, almost all the engine oil will have drained into the sump, leaving the crankshaft bearings, cylinder walls and other surfaces virtually dry; when starting the engine in this condition there is a danger of scuffing if it runs for a few seconds before the oil has circulated to all the potential wearing points. This undesirable state of affairs can usually be avoided on these cars as the engines will rarely start from cold without the use of the choke, and so it is a simple matter to allow the engine to turn over a few times on the starter until the oil pressure warning light is extinguished, before pulling out the choke control. This precaution is not, of course, necessary for the everyday cold start, where a film of oil still exists around the engine.

Thanks to the vast improvements made in tyres in recent years, the general road behaviour of the big Fords, which set high standards in the 1950s, can be further enhanced today. Those enthusiasts who wish to continue to use cross-ply tyres, in the interests of originality, unfortunately now have only a limited choice. The Goodyear G8 and the Lee Conquest (also a Goodyear product) are available in the 6.40 × 13 size to suit Zephyrs and the 6.00 × 13, which can be used on Consuls instead of the slightly smaller, and no longer available, original size. Both these tyres offer qualities of adhesion far superior to the original-equipment cross-ply tyres of 20 or 30 years ago, so driving on wet roads should present no cause for alarm with a car thus shod. When using cross-plies on the Zephyr, inflating the front tyres to a pressure 4psi greater than the rears lessens the basic understeer, giving almost neutral handling characteristics without impairing the car's inherently good directional stability. This pressure differential is preferred by many Zephyr owners.

There is a wide choice of radial-ply tyres available in suitable sizes. Textile radials will usually tend to give heavier steering coupled with a distinctly poorer ride when fitted to these models, and steel radials are much more suitable. For both the Mk 1 and Mk 2 Zephyr, Michelin recommend the 175 SR-13 size from their MX range. With equal front and rear pressures the radials will lessen the basic understeering tendency of these cars, but this can be restored if desired by choosing rear pressures higher than at the front. Michelin recommend 25psi front and 28psi rear for normal motoring, increased to 31psi front and 34psi rear for high-speed driving. The enthusiastic Zephyr driver who prefers neutral handling characteristics should find that pressures of 32psi all round with steel radials will give excellent results whilst retaining the normally light steering.

Michelin recommend their 165 SR-13 MX steel radial for the Consul. At the pressures of 28psi all round specified in the Consul's handbook, radials may give slightly heavier steering, but this effect will not really be noticeable at much above parking speeds. Because the Consul is lighter than the Zephyr and running on smaller tyres, experimenting with pressure differentials to alter the handling characteristics is not to be advised; any significant increase in front pressures can lead to adhesion problems under some circumstances.

Whilst supplies for routine servicing and the choice of expendable items like tyres will concern the owner of a car in running order, the need for major overhaul presents another set of problems. There is considerable intechangeability of parts between the principal mechanical components of Mk 1 and Mk 2 models, and the engines of the subsequent Mk 3 range share the family affinity, too. This is of great value when spares are required for overhaul, and some notes on possible swaps and substitutions may well be useful.

Because of the differing cubic capacities, the piston and ring assemblies, conrods, crankshaft and its bearings are not interchangeable between Mk 1 and Mk 2, although the big-end and main bearing bolts are the Mk 1 item in both cases. The Mk 3 pistons, rings and conrods, (211E Part Nos) can be used in sets in the Mk 2 engines, and the Mk 3 Zephyr 4 crankshaft (211E 6303-B) is interchangeable with that of the Mk 2 Consul. The Mk 1 Consul/Zephyr/Zodiac and the Mk 2 Zephyr/Zodiac use the same oil pump (EOTA 6600-C) and this is interchangeable with the Mk 3 Zephyr 6/ Zodiac assembly (213E 6600). The oil pump in the Mk 2 Consul (204E 6600) can be replaced with that of the Mk 3 Zephyr 4 (211E 6600). The timing case cover and gasket are common to Mk 1 and Mk2, as are the front and rear crankshaft oil seals, although the front seal can also be replaced by the Mk 3 seal (211E 6700). The Mk2 and Mk 3 cylinder head gaskets are interchangeable. The valve gear for both Mk 1 and Mk 2 is almost totally standard Mk 1. The camshaft, camshaft bearings, cam followers, pushrods, rocker shaft, rocker arms, valve springs and standard-size valves are all interchangeable and have Mk 1 (EOTA/EOTTA) Part Nos. The Mk 3 camshaft bearings (211E 6262-E) and cam followers (211E 6489-A) can also be used in place of the original (EOTA) items. The later camshafts (EOTA/EOTTA 6250-B) fit both Mk 1 and Mk 2, but the original camshafts without the B suffix fit the Mk 1 only. The timing chain (E93A 6270) on both models is from the earlier 10hp side-valve engine.

On the Mk 1 Consul, the clutch can be replaced with the heavier-duty Mk 1 Zephyr assembly, but this is not possible on the Mk 2 models because the Mk 2 Zephyr has a larger diameter clutch than the Mk 2 Consul. The clutch release bearing (E72 7580-A) is the same for all the Mk 1 and Mk 2 cars, but the bearing hub (EOA 7561) Mk 1 and (204E 7561) Mk 2 differs slightly between the two models and is not interchangeable. Except for the main shaft assembly (E 23 GA 2) Mk 1 and (E 40 GA 2) early Mk 2, or E 410 GA 2) Mk 2 from January 1961, virtually all the gearbox internals are the same for both models. The original circlips (EOA & EOTTA Nos) can be replaced with the later improved Mk 2 circlips (204E Nos) except those (EOTTA Nos) that are applicable to the overdrive unit only. When overhauling an early Mk 2 gearbox the later mainshaft assembly can be used to replace the original.

Although of basically the same design, because of the difference in size and weight between the two models, the steering gear, springs and suspension units are not interchangeable. However, the front suspension unit thrust bearing (EOA 5426-B) and its circlip (EOA 5430) are the Mk 1 items in both models, and the worm-and-peg steering box in the pre-October 1957 Mk 2 cars utilizes the Mk 1 peg (EOA 3566-A).

The starter motors, (EOTA 11001-B) Mk 1, (204E 11001-B) Mk 2, and (100E 11001-C) Anglia/Prefect, are all interchangeable. The Mk 2 model prior to February 1959 also uses the Mk 1 generator (EOTA 10001-B), whereas the later Mk 2 has the Anglia generator (105E 10001-B).

Driving a Consul, Zephyr or Zodiac

To those people who have not experienced the type, the thought of a three-speed gearbox with an unsynchronized first gear, and controlled by a column-mounted lever, is off-putting to say the least. Yet the three-speed column change fitted to the big 1950s Fords is one of the features which helps to give the 'character' that many enthusiasts recognize in these models. Unlike some of the four-speed-equipped cars of the period, where first gear was really an emergency low ratio and second could be used for starting on the level if the driver so wished, the first gear ratio on the big Fords is really quite high, and at times its selection is needed when the car is on the move. The method used to engage this unsynchronized first gear whilst the car is in motion is known as double declutching, a technique that in the prewar years was standard practice for most drivers, but which has largely disappeared over the last 25 years or so with the advent of all-synchromesh gearboxes. With the car travelling in second gear, the clutch pedal is depressed and the gearlever is moved downwards into the neutral position. The clutch pedal is then quickly released and the accelerator is opened in order to increase the engine revs to the speed at which it would be running if first gear were already engaged. As the accelerator is released again, the clutch is quickly depressed and the gearlever moved across and down into the first gear position. With the gear now engaged the drive is taken up again as with any normal downward change. Matching the engine revs exactly to the speed is the difficult part, and the secret here is to have a few more, rather than less, revs, so that as the accelerator is released whilst the gearlever is brought close to the first position, the revs will be falling and at the point when engine speed and road speed coincide the lever will slide easily into the first gear position. Once mastered, this technique can be a source of some satisfaction as it is now one of the almost-lost 'arts' of driving. To some drivers the three-speed gearbox actually seems advantageous; for example, it is that much easier to 'be in the right gear at the right time', as the driving instructors always used to say, if there is only one intermediate ratio rather than two to choose from.

Above all, it is important to accept the three-speed gearbox for what it is and drive accordingly. The indirect ratios used in the Consul and Zephyr box are such that second gear falls somewhere between the usual second and third ratios in a four-speed unit, and so the gap that exists between second and top gear is far less than would be the case if third gear were removed from a typical four-speed arrangement. In other words, the driver used to four-speed boxes should view second gear on these cars as a rather low third in some circumstances, as well as being a very useful and wide-ranging second. The flexibility of these engines plays an important part in overcoming the problem which might otherwise be apparent with only one intermediate gear available. Even the four-cylinder versions can be considered primarily as top-gear performers, as they will pull very well indeed above about 25mph in this ratio. If and when a rather more vigorous effort is called for, for example when overtaking a slow-moving truck, then second gear really comes into its own, providing a better rate of acceleration than top up to around 42mph in the Mk 1 Consul and about 47mph with the Mk 2. Although the engines will quite willingly rev beyond these speeds in second gear by about another 5 or 8mph, respectively, they are then going well over the peak of the power curve and acceleration is tailing off. The corresponding useful speeds in first gear are about 23mph and 27mph for the Mk 1 and Mk 2, respectively, these speeds being some 5 or 6mph below the absolute maxima in this ratio.

The three-speed gearbox makes most sense when mated to the big and extremely flexible six-cylinder engines of the Zephyrs and Zodiacs, in which many miles can be covered in give-and-take conditions without the need to touch the gearlever. If, for instance, speed has been checked to 20mph or so and then the situation clears, unless 40mph needs to be attained in less than 8 seconds there is no real point in employing second gear, and those drivers who habitually make full use of this flexibility usually find that these cars are surprisingly economical by six-cylinder standards. Once again the maximum speeds possible in the indirect gears are far higher than is actually needed for optimum performance, and when accelerating hard from a standing start, changes at around 26mph and 44mph in the Mk 1 Zephyr should produce the best results. The corresponding figures for the Mk 2 Zephyr are a rather convenient 30mph and 50mph.

When maximum acceleration is needed, perhaps in a tight overtaking manoeuvre, it is worth noting that the best results are obtained if the accelerator is opened progressively towards the fully open position. The accelerator pump incorporated in the Zenith carburettor provides an enriched mixture whilst the accelerator pedal is being opened, and if the pedal is simply snapped open to the full position the extra fuel provided will be used up in that instant and almost completely wasted. Depending on the position of the accelerator pedal when the manoeuvre is started, something between 2 to 3 seconds taken to reach the fully open position will provide a quicker rate of progress than the sudden full-throttle approach.

For any of these cars to give their best in the handling department it is important that the springs and shock absorbers are in good condition; if these items have weakened, then replacement will be necessary before the full cornering capabilities of the model can be achieved. In order to enjoy the good cornering powers these cars possess it is also important to know that they respond best to what always used to be accepted as the 'correct' technique, with all the decelerating and braking done on the straight and the corner taken with the accelerator open and the engine thus driving the car round. Although it has only a four-cylinder engine, the weight distribution on the Mk 1 Consul is very definitely nose-heavy, and gives a consistent amount of understeer. With the relatively small amount of power available, even wide throttle openings will not unduly lessen this tendency, so fast cornering on this model is simply a case of keeping the power on and using the steering to hold the chosen line. On the Mk 2 Consul the picture changes slightly; this car has the most even weight distribution of the entire range and consequently it understeers rather less than the Mk 1 Consul. This factor, coupled with the slightly larger capacity engine, means that under some circumstances it is possible to have almost neutral steering characteristics with the power on in a corner, and this enables the Mk 2 Consul to hold the chosen line that little bit more easily and at a somewhat higher speed than is possible with the earlier car.

With the relatively heavy six-cylinder engine up front, the Zephyrs and Zodiacs enjoy excellent directional stability and a consistent degree of understeer during cornering. With the ample power available, this understeer can be counteracted up to a point by progressively increasing the throttle opening as the car goes through the bend, and if this technique is accurately employed the steering lock can be wound off after the apex of the corner, enabling an excellent line to be taken out of the bend with the car accelerating strongly. With the raised front, normal rear, tyre pressure differential that has been mentioned earlier as being effective when the Zephyrs are running on cross-ply covers, slight oversteer can sometimes be detected when the progressive power-on technique is being used, and if this is felt to be undesirable, simply easing up on the accelerator will be sufficient to restore the neutrality. To get the very best results from the Mk 1 Zephyr, a considerable degree of accuracy is required when setting up the car for a bend, and for fast road work this car will always respond best if excess speed is removed before entry and is then followed up with the progressively opening throttle method through the turn. If, however, this car is taken into a bend much too fast on the overrun, it can become untidy, although thanks to the good adhesive qualities of modern well-known brands of tyre it should not become the 'handful' that it was sometimes said to be in its early days. It is a little easier to get the best results from the Mk 2 Zephyr, and when handled really well on twisting roads this car is capable of maintaining very quick progress. Unlike the earlier model, the Mk 2 is also quite forgiving when driven badly, and when a corner is entered a little too fast on the overrun it will simply lean over, dig in the outside front wheel and scrub its way round, whilst usually still managing to hold a reasonably good line.

CHAPTER 7

A very special Zephyr

Case history of Peter Scott's car

Many excellent and interesting examples of the big Fords from the 1950s have come to light since the formation of the owners' clubs in 1977, but there is surely none more remarkable than RUR 5, which has been owned by Peter Scott, of Barnet in Hertfordshire, ever since it was new. Riding in a friend's model T Ford in the early-1920s had triggered off in Peter Scott an interest in Fords that would continue to grow enthusiastically over the years. As his own motoring experiences broadened he began to visualize a car that would suit his particular requirements right down to the last detail. A succession of Fords, which had included one of the original Y-type Populars, and a secondhand 1935 10hp model still affectionately remembered as 'Barrel Billy', had given him good service, and being a gifted amateur engineer himself he had always appreciated their sound technical features and the quality of the workmanship and materials which had gone into their construction. So it was becoming evident that the 'ideal' car that he had in mind would inevitably be based on a Dagenham product.

What he envisaged was a high-performance saloon that would easily swallow up two people, a large inflatable boat complete with its outboard motor, and all the other luggage and paraphernalia that would be needed on long continental tours. A maximum speed of 100mph would be desirable so that even with all the gear on board an easy 80mph cruise could be maintained along the German autobahns, and acceleration, hill-climbing and handling would have to be good enough to allow Alpine passes to be stormed in something very close to rallying style. At the £1,000 or so which he had available in 1953 there was nothing in the showrooms to fit the bill exactly, but the well-proven, Monte Carlo Rally-winning Zephyr Six, at a tax paid price of £754, certainly did seem to have the potential.

An order was placed with Dagenham Motors, and in March 1954 a brand new Zephyr Six, registration number RUR 5, was collected from their London garage and driven home through a heavy thunderstorm to Barnet. The first stage in achieving that 100mph capability was now begun. It would have been relatively easy to have bought some extra performance by purchasing a tuning conversion, such as those offered by Raymond Mays or Laystall, but Peter was keen to do the cylinder head work himself. The standard head was first exchanged for the optional 7.5:1 compression

A recent picture of Peter Scott and RUR 5, the inseparable pair who have travelled more than 300,000 miles together. Photo from Peter Scott.

ratio Zodiac fitting; this was then gas-flowed by Peter himself in the home workshop at Barnet. The valve ports were carefully opened out and polished, and on reassembly the original valve springs were replaced with Terrys Aero double springs. To get the gases away a complete Servais exhaust system was fitted, comprising a six-branch manifold, a 'straight through' cylindrical silencer 4ft long by 6in diameter, and a large-bore tail-pipe. To give a fatter spark at high rpm a Lucas Sports coil was substituted for the original. An auxiliary electric fan was added to the cooling system along with an automatic tube and bottle radiator replenishing system, whilst warming up was to be assisted by a radiator blind mechanically operated by a crank handle beneath the dashboard in front of the driver.

In view of the heavy loads and high speeds anticipated, Firestone Nylon Sports tyres of the standard 6.40 × 13 size were chosen. Uprated front springs, obtained from the Ford Competitions Department, and Thames van rear shock absorbers were also fitted to help to take care of the load, whilst axle tramp was eliminated by Scott-designed and made nylon-bushed anti-tramp bars.

A very large, heavy-duty Exide battery was found a home in the luggage boot on the opposite side to the spare wheel. All the electrical circuits were individually earthed and fused, the fuseboard being situated on the interior side of the dashboard. The radio aerial, fuel filler cap and windscreen washers are all electrically operated, the washers in particular reflecting Peter Scott's ingenuity; the water line runs inside the windscreen rubber beading to two outlets that are directly above the wipers, and thus allow the water to flow downwards over the screen and on to the wiper blades. A well-stocked instrument panel, designed and made by Peter himself, and carrying fuel gauge, ammeter, combined oil pressure and water temperature gauge, rev-counter, speedometer and clock, sits neatly on top of the steering column shroud in place of the original Ford speedometer housing.

Before running-in was commenced, attention was turned to protecting the bodywork. All the body joints and seams were treated with Dum Dum

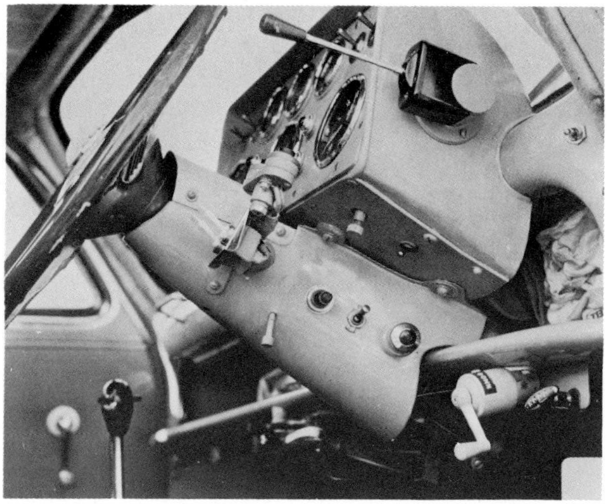

Cockpit layout of RUR 5, showing Scott-designed and manufactured instrument panel and remote, floor-mounted gear-change. The crank handle under the parcels shelf is for the radiator blind. Photos by Peter Scott.

waterproof putty, accessibility for this operation being made easy by the bolt-on construction of the front panel and all four wings. The entire under-surfaces were then coated with a home-made mix of lead paint and rubber solution. By now some £200 had been spent on the various bits and pieces, in addition to around 600 hours' work, and RUR 5 was ready for the road. Part of the running-in procedure was accomplished by going from Barnet to Land's End and back on two consecutive weekends.

At this stage, RUR 5 was already considerably quicker than a standard Zephyr, and the suspension modifications ensured that handling and road clinging could match the extra urge. With the brakes, however, it was a somewhat different story, and stopping the Zephyr at an all-up weight with the boat on board of around 3,500lb from speeds approaching 90mph would be a dicey business. The only modifications to the braking system so far had been the introduction of automatic brake bleeders to make fluid changes a simple, one-man operation and thus ease the servicing situation. A set of Ferodo/Girling bonded shoes with a harder than standard grade of lining material, aided by a servo unit which was salvaged from a crashed Connaught racing car, were then installed, and thereafter RUR 5 could stop as well as it could go. The servo unit fitted neatly under the bonnet in the space that had been vacated by the battery.

In the quest for more speed, experiments were carried out with both twin and triple SU carburettor arrangements, but Peter has always had a preference for the simplicity of a single instrument, and a modified and rejetted Zenith, aided by a forward-facing air intake situated under the front of the car and feeding directly into the air cleaner, was found to give very satisfactory results.

In 1957, two further modifications were carried out which put RUR 5 into the genuine 100mph category; the standard crownwheel-and-pinion assembly of 4.444:1 ratio was replaced by a 3.90:1 unit obtained from the Ford Competitions Department, and a Laycock de Normanville overdrive was fitted to the standard three-speed gearbox. The electrically-operated overdrive is controlled by a switch in the top of the gearlever, which is floor-mounted on RUR 5, and is yet another example of Peter Scott's skill, being designed, manufactured and fitted in the home garage in Barnet while the gearbox was out of the car for the installation of the overdrive unit. RUR 5 could now manage just over 100mph in overdrive top and would give a fuel consumption of 22-23mpg cruising at 80mph. In this form the car was taken on a 4,800-mile tour of central Europe during the 1957 summer.

By 1960, the engine, and indeed the whole car, were in peak form, and in June of that year, with some 70,000 miles already behind it, RUR 5 was pounding across Thetford Heath, in Norfolk, in a gale-force wind that was following precisely the same direction. The speedometer quickly left the 100mph scale and settled against the stop, whilst the rev-counter carried on enthusiastically to 6,800rpm which, even allowing for some optimism in the instrument, equated to a speed in excess of 120mph.

The long continental tours each summer, at high speeds and with a heavy load on board, inevitably took their toll, and at 110,000 miles Peter decided that wear in the engine was sufficient to warrant a complete overhaul. The unit was removed from the car and dismantled. The cylinder block was rebored, and the new pistons that were to be fitted were carefully balanced before reassembly. The reground crankshaft was balanced by Alexander, the tuning specialists, and the cylinder head was skimmed to raise the compression ratio still further, the new figure of 9.7:1 being about the limit that can be achieved with the original cylinder head. The power output of the engine has never been measured, but taking into account the original modifications, coupled with the balanced rebuild, a figure of 95bhp would seem to be a reasonable assumption at this stage in the career of RUR 5. In 1966, after a particularly fierce 'pass storming' holiday in the Alps, which had involved much full-throttle first and second gear work, some trouble arose with the gearbox; as the mileage had by now exceeded 120,000, an overhaul of this unit seemed wise and was undertaken by Peter himself.

Today, more than 28 years after first taking to the road, and with some 315,000 miles having passed beneath its wheels, RUR 5 still runs as well as ever. There have inevitably been some other repairs and replacements along the way. The car is now on its third clutch assembly, and its fourth Servais 'straight through' silencer was fitted in 1980. The fuel pump and water pump are both replacement items, in 1970 the radiator was reconditioned,and more recently the starter motor was overhauled. A new brake master cylinder has been fitted, and all the brake wheel cylinder seals have been renewed. Brake lining wear has been moderate, thanks no doubt to the harder grade material, and the car is on only its third set of shoes.

Balancing the engine components at the rebuild has certainly paid dividends; in the 200,000-plus miles that have since been covered the only engine replacements have been the timing chain and another set of valve springs. Oil changes at 2,500-mile intervals have also played their part in the low rate of engine wear. Castrol is always used — XXL in the early days and latterly GTX. The benefits gained from regular attention with a grease gun, loaded in this case with Duckhams Laminoid, are amply demonstrated by the fact that apart from a new idler arm assembly, fitted in 1979, the steering gear is still original, and the universal joints on the propeller shaft have also survived from new.

Tyre pressures of 30psi front and 26psi rear, fully loaded (two people and boat on board), or 2psi less all round solo, have always been meticulously maintained, with the result that RUR 5 is only now on its fifth set of covers. The home-made underseal has been periodically inspected and reapplied wherever and whenever necessary, and consequently the whole structure of the car is still virtually perfect. Regular washing and a twice-yearly polishing of the paintwork and chrome-plating with 'old fashioned' Simoniz have endowed RUR 5 today with a remarkably well

The additional brake fluid reservoir, six-branch manifold and brake servo in the nearside front are all evident in these views. Chrome-plated rocker cover and air cleaner enhance the underbonnet scene. Photos by Peter Scott.

kept appearance, although the pleasing Opal colour does seem to have faded a little in places, perhaps because of all those hours of exposure to the brilliant continental sunshine.

Naturally enough, in a life of more than 300,000 miles that has taken RUR 5 across France, West Germany, Holland, Belgium, Denmark, Italy, Monaco, Switzerland and Austria, many times over, there have been some incidents, but fortunately no major accidents. Collision with a pheasant, which was taken home and served for dinner, necessitated the fitting of a new windscreen. Far more alarming was the occasion when both front tyres blew out almost simultaneously at high speed on the autobahn. 'It was a very anxious few moments, but no real disaster', recalls Peter. The German ADAC, an organization similar to the RAC in Britain, came to the rescue on this occasion with a pair of suitable tyres obtained in Munich. In France in 1972, RUR 5 was brought to another involuntary stop after being filled up with petrol that proved to be contaminated with water, and a considerable delay ensued while the fuel lines were disconnected and the system drained and replenished. The comprehensive tool kit that is stowed permanently in RUR 5 was brought into service this time. Peter also remembers, with some amusement, being stopped by Police in Italy for theoretical non-observance of local overtaking rules on an autostrada, and having a furious row when offering a 10,000-lire note for a 2,000-lire fine — they had no change!

Peter Scott will never sell RUR 5; instead, when he, to use his own phrase, 'stops living', the old Zephyr will also return to its makers under the provisions of a clause already included in his will — 'To the Ford Motor Company, for their museum, my Mark 1 Zephyr car, RUR 5, chassis and engine number 80,508, as some appreciation of its excellence and the pleasure it has given me'.

CHAPTER 8

The Ford-powered sports cars

Allard, Lea-Francis, Reliant and others

Builders of specialist, limited-production sports cars, seeking to create something more exclusive than run-of-the-mill transport and exploiting areas of the market too small to interest the major manufacturers, have often chosen Ford engines because of the combination of moderate cost, widely available spares and suitability for tuning which they offer. The earlier side-valve units, 8, 10 and V8, had been popular in this role, and the new generation of OHV engines soon attracted the same kind of attention.

The first car of this type to utilize the new Consul/Zephyr engine was the Allard Palm Beach. Announced in August 1952, the Palm Beach featured a tubular-steel chassis framework which was designed to house as many as possible of the Consul/Zephyr components, the radiator, engine, clutch, gearbox and back axle all being the standard Ford Mk 1 items, as were the brakes and the 13in road wheels, complete with Mk 1 hubcaps, although optional 15in wire wheels could be specified. A Zephyr steering wheel, complete with horn ring, operated Marles steering gear, whilst Allard-designed coil-spring suspension, independent at the front and with radius arms and a Panhard rod locating the rear axle, completed the running gear specification. All this was clothed in a very clean, full-width aluminium body, and the cockpit featured a leather-upholstered bench seat which, in conjunction with an offset floor-mounted gear-change, made this a three-seater sports car.

As a sports car, the Consul-engined version was underpowered, and although a twin-carburettor, high compression ratio option was available, only eight of the four-cylinder cars were in fact produced. Much more attractive, at a price of £1,063, was the Zephyr-engined version, which was able to reach 70mph from rest in a little over 20 seconds on its way to a maximum speed nudging 90mph, figures which, of course, could be considerably improved upon if the owner was prepared to go to the additional expense of an engine conversion such as that which Raymond Mays was offering. Production of the Zephyr-engined Palm Beach had totalled 65 when the model went into Mk 2 form in 1956, after which it was available to special order only; just seven of the Mk 2 models were made. According to the Allard Owners' Club, 12 examples of the Palm Beach have survived, a figure which makes this car a very exclusive property today.

A full four-seater convertible, known as the Paramount Roadster,

Allard Palm Beach. Pictured here is one of the rare Mk 2 models with the 2,553cc Mk 2 Zephyr engine. The aircraft in the background, a De Haviland Mosquito, may well also have Ford connections, as the Rolls-Royce Merlin aero engine was mass-produced by Ford of Britain during World War 2. Photo from Patricia Hulse.

powered by a Ford 10hp side-valve engine, had seen limited production in the early-1950s, but had proved to be seriously underpowered, and in an effort to rectify this shortcoming, Paramount Cars offered the Mk 1 Consul engine in the Roadster. Although featuring an aluminium body, the Paramount was constructed in a rather traditional manner, which resulted in much excess weight, and consequently performance which could not match the attractive sporting appearance which the model possessed. At £1,013, the Paramount was some £200 more expensive than the slightly smaller and faster Consul convertible, whose engine it used, and so it never really stood a chance of significant sales success.

1959 saw the introduction of the Fairthorpe Zeta, a Zephyr-engined sports car which could be bought in kit form for home assembly by the practical enthusiast, in addition to being available for purchase in fully-assembled configuration at a price of £1,408. The engine was offered with the Raymond Mays light-alloy cylinder head and either three SU or six Amal carburettors. With its two-seater glass-fibre bodywork, in conjunction with a lightweight square-section tubular-steel ladder-type chassis, keeping the unladen weight down to less than 14cwt, the Zeta was a very accelerative car, and Girling disc brakes were employed to tame the power when necessary. The Zeta was successfully raced and rallied by Air Vice Marshal D.C.T. 'Pathfinder' Bennett, who was associated with the Fairthorpe company.

After an absence of several years, the famous name of Lea-Francis was back at Earls Court in 1960, with the prototype of a new $2\frac{1}{2}$-litre roadster, the Lea-Francis Lynx. The engine chosen was the Ford six-cylinder unit, modified in this instance to give a power output of 107bhp at 4,500rpm, the modifications being a raised compression ratio of 8.0:1, three SU carburettors, each on a separate induction pipe, and a two-branch exhaust manifold. The power was transmitted to the live rear axle via a Moss four-

The sole production model Lea-Francis Lynx, now the property of Lea-Francis enthusiast Keith Tricker, who also owns the two prototype cars. Photo by Keith Tricker.

speed gearbox with overdrive, whilst the running gear comprised torsion bar independent front suspension, Lea-Francis designed rack-and-pinion steering, leaf springs at the rear and Dunlop disc brakes all round. The futuristically styled occasional four-seater bodywork was in steel and aluminium, and was attached at 10 points to an extremely robust tubular-steel ladder chassis. After the Motor Show appearance, the prototype Lynx was dismantled and development of the model was completed on two further cars. Additional modifications to the engine consisted of a further increase in compression ratio, larger valves, a high-lift camshaft and a full dual exhaust system, the result of which was to raise the power output to 129bhp. The third car built was regarded as a production model and was sold as such, but unfortunately, Lea-Francis went into liquidation soon after this, and development of the Lynx concept was brought to a premature end. Lea-Francis enthusiast Keith Tricker has managed to acquire all three Lynxes, although all that remained of the prototype when he finally tracked it down in a scrapyard was the rusty bodyshell. Lynx No 2 still awaits restoration, whilst No 3 is restored and in full running order, its tuned Zephyr engine giving it, according to Mr Tricker, a 0-60mph capability of around 10 seconds, and an estimated maximum speed of 115mph in overdrive, coupled to a useful fuel economy average of 26mpg.

Another car to feature the Zephyr engine and Moss gearbox combination was the Ruddspeed Ace. This car was the already well-known AC Ace, which was normally available with either a single-OHC 2-litre AC engine, or the crossover-pushrod 2-litre Bristol unit. K.N. Rudd (Engineers) Ltd carried out the conversion to Zephyr power, and were able to offer the Ace with the standard Zephyr unit at £1,678, which represented a saving of more than £100 on the AC-engined version, and a colossal £450 saving over the Bristol-powered car. In addition to the standard Ford engine, Ken Rudd offered several different stages of tune, based on both the standard cast-iron cylinder head and the Raymond Mays light-alloy component, with either twin or triple SUs or triple Webers. Special sets of lightweight pistons and pushrods were also available with any of the tuning options, and these offered a further 5bhp increase. An

The Reliant Sabre, the Consul-engined forerunner of the successful Scimitar coupe and GTE models. Photo from the Reliant Motor Company.

Ace with the Mays head, triple SUs and the lightweight pistons and pushrods was sampled by *The Motor*, and in their hands it accelerated from rest to 100mph in 23.3 seconds, and achieved a maximum speed of 116mph. A total of 38 examples of the Ruddspeed Ace were produced, and the Zephyr engine also found its way into eight examples of the closed coupe version, the Aceca. Apart from one which was destroyed in an accident, all the Ruddspeed cars are thought to have survived.

The Reliant Engineering Company, famous for their utilitarian three-wheeler economy cars, took their first steps into the specialist sports car market late in 1961 with the introduction of the Sabre. The Mk 2 Consul engine employed retained its single Zenith carburettor, although it did have a modified cylinder head, which raised the power output to approximately 70bhp, and in conjunction with a German ZF gearbox this was quite sufficient to propel the glass-fibre-bodied two-seater from rest to 70mph in around 20 seconds and on to a 90mph maximum speed. A somewhat unusual divided-axle front suspension aroused comment at the time, but in practice it proved to be perfectly satisfactory and the Sabre earned a good reputation as a distinctive small sports car, although in its early days few people could have realized just how significant this model was for the Reliant company. In 1962, the design was extensively reworked to take the six-cylinder Ford engine, which was by now in Mk 3 form, and this led in turn to the introduction in 1964 of the elegant Scimitar GT Coupe, the forerunner of the Scimitar GTE models which remain in production today.

In the big Ford's export stronghold of Australia there were several attempts to produce glass-fibre-bodied sports cars during the 1950s, some of which were planned around Consul/Zephyr engines and running gear. A particularly neat example was designed by Nat Buchanan, and was offered with either Consul or Zephyr multiple-carburettor engines accompanied by the standard three-speed gearbox, but with floor-mounted change. Buckle Cars Ltd, of Sydney, produced 22 examples of their Zephyr-based sports car, the early versions of which incorporated many Mk 1 components, with the gearbox once again featuring a floor-change conversion. The specification was updated in 1957 to include the Mk 2 Zephyr engine, although items such as the brakes, steering box and wheels remained Mk 1. At least one example of the Mk 2 version has survived, and was recently subjected to an extensive restoration programme by its owner, Ian Clayton.

APPENDIX A
Technical specifications

CONSUL (Mk 1)
Engine: 4-cylinder in-line, overhead valves. Bore × stroke 79.37mm × 76.2mm, capacity 1,508cc. Compression ratio 6.8:1. Maximum power 47bhp (nett) at 4,400rpm. Maximum torque 72lb/ft (nett) at 2,400rpm.
Transmission: 3-speed manual gearbox. Axle ratio 4.625:1 (4.556:1 from November 1952). Mph/1,000rpm with 4.625:1 axle: 1st 5.24, 2nd 9.06, top 14.9; with 4.556:1 axle: 1st 5.38*, 2nd 9.3*, top 15.3 (*lower 1st and 2nd gear ratios fitted between November 1952 and April 1953 giving 4.67 and 9.04mph/1,000rpm, respectively).
Tyres: Cross-ply, 5.90 × 13, pressure 28psi. When normal-profile 165—13 radial-ply tyres fitted, overall gearing is reduced by approximately 3%.
Brakes: Lining area 121sq in.
Steering: Ratio 13.6:1, $2\frac{1}{2}$ turns lock to lock. Turning circle 40ft 6in.
Dimensions: Length 13ft 10in, width 5ft 4in, height 5ft $0\frac{3}{4}$in, wheelbase 8ft 4in, ground clearance $6\frac{1}{2}$in, track 4ft 2in front, 4ft 1in rear.
Weight: $21\frac{3}{4}$cwt unladen.

ZEPHYR SIX and ZODIAC (Mk 1)
Engine: 6-cylinder in-line, overhead valves. Bore × stroke 79.37mm × 76.2mm, capacity 2,262cc. Compression ratio 6.8:1 (Zodiac 7.5:1). Maximum power 68bhp (nett) at 4,000rpm (Zodiac 71bhp at 4,200rpm). Maximum torque 108lb/ft (nett) at 2,000rpm (Zodiac 112lb/ft at 2,000rpm).
Transmission: 3-speed manual gearbox. Axle ratio 4.375:1 (4.444:1 from November 1952). Mph/1,000rpm with 4.375:1 axle: 1st 5.75, 2nd 9.95, top 16.35; with 4.444:1 axle: 1st 5.68*, 2nd 9.83*, top 16.15 (*lower 1st and 2nd gear ratios fitted between November 1952 and April 1953 giving 4.93 and 9.55mph/1,000rpm, respectively). Mph/1,000rpm with 4.444:1 axle and Borg-Warner overdrive: 2nd 14.2, top 23.0; with Laycock overdrive: 2nd 12.8, top 20.7.
Tyres: Cross-ply, 6.40 × 13, pressure 24psi. When normal-profile 175—13 radial-ply tyres fitted, overall gearing is reduced by approximately 3%.
Brakes: Lining area 121sq in.
Steering: Ratio 13.6:1, $2\frac{1}{2}$ turns lock to lock. Turning circle 41ft 6in.
Dimensions: Length 14ft $3\frac{3}{4}$in, width 5ft 4in, height 5ft $0\frac{3}{4}$in, wheelbase 8ft 8in, ground clearance 7in, track 4ft 2in front, 4ft 1in rear.
Weight: $23\frac{1}{4}$cwt (Zodiac $23\frac{3}{4}$cwt) unladen.

CONSUL (Mk 2)
Engine: 4-cylinder in-line, overhead valves. Bore × stroke 82.55mm × 79.5mm, capacity 1,703cc. Compression ratio 7.8:1 (6.9:1 optional). Maximum power 59bhp (nett) at 4,400rpm (55bhp at 4,400rpm with 6.9:1 CR). Maximum torque 91lb/ft (nett) at 2,300rpm (87lb/ft at 2,300rpm with 6.9:1 CR).
Transmission: 3-speed manual gearbox. Axle ratio 4.11:1. Mph/1,000rpm: 1st 5.95, 2nd 10.2, top 16.9.
Tyres: Cross-ply, 5.90 × 13, pressure 28psi. When normal-profile 165—13 radial-ply tyres fitted, overall gearing is reduced by approximately 3%.
Brakes: Lining area 147sq in. With disc/drum system, swept area 299sq in.
Steering: Ratio 16.8:1, $3\frac{1}{4}$ turns lock to lock (ratio 18.1:1, $3\frac{1}{2}$ turns lock to lock, from October 1957). Turning circle 34ft.
Dimensions: Length 14ft 4in (14ft $6\frac{1}{2}$in from February 1959), width 5ft 9in, height 5ft 2in (5ft $0\frac{1}{2}$in from February 1959), wheelbase 8ft $8\frac{1}{2}$in, ground clearance $6\frac{1}{4}$in, track 4ft 5in front, 4ft 4in rear.
Weight: $22\frac{1}{4}$cwt unladen.

ZEPHYR and ZODIAC (Mk 2)
Engine: 6-cylinder in-line, overhead valves. Bore × stroke 82.55mm × 79.5mm, capacity 2,553cc. Compression ratio 7.8:1 (6.9:1 optional). Maximum power 85bhp (nett) at 4,400rpm (81bhp at 4,400rpm with 6.9:1 CR). Maximum torque 133lb/ft (nett) at 2,000rpm (127lb/ft at 2,000rpm with 6.9:1 CR).
Transmission: 3-speed manual gearbox (automatic transmission optional extra). Axle ratio 3.90:1. Mph/1,000rpm: 1st 6.5, 2nd 11.27, top 18.5. With Borg-Warner overdrive: 1st 9.2, 2nd 16.1, top 26.42. With Laycock overdrive: 2nd 14.52, top 23.99.
Tyres: Cross-ply, 6.40 × 13 (6.70 × 13 for estate car and with automatic transmission), pressure 24psi. When normal-profile 175—13 radial-ply tyres fitted, overall gearing is reduced by approximately 3%. 6.70 × 13 tyres (estates/automatics) raise overall gearing to 19.1mph/1,000rpm.
Brakes: Lining area 147sq in. With disc/drum system, swept area 299sq in.
Steering: Ratio 16.8:1, $3\frac{1}{4}$ turns lock to lock (ratio 18.1:1, $3\frac{1}{2}$ turns lock to lock, from October 1957). Turning circle 36ft.
Dimensions: Length 14ft $10\frac{1}{2}$in (Zodiac 15ft $0\frac{1}{2}$in) prior to February 1959, 14ft 11in (Zodiac 15ft $0\frac{1}{2}$in) from February 1959, width 5ft 9in, height 5ft 2in (5ft $0\frac{1}{2}$in from February 1959), wheelbase 8ft 11in, ground clearance $6\frac{1}{4}$in, track 4ft 5in front, 4ft 4in rear.
Weight: 24cwt (Zodiac $24\frac{1}{2}$cwt) unladen.

APPENDIX B
Colour schemes

Mk 1 CONSUL and ZEPHYR SIX
Saloons
Winchester Blue
Westminster Blue
Channel Green
Deep Carriage Green
Ivory*
Metallic Light Blue**
Metallic Light Olive**
Metallic Light Coral**
Edinburgh Green
Lichfield Green
Canterbury Green
Dorchester Grey
Bristol Fawn
Opal Blue**
Essex Blue**
Honey Beige**
(* Export only)
(** Australia/New Zealand only)

Convertibles

Westminster Blue	Black
Winchester Blue	Mandarin Red*
Canterbury Green	Ivory*
Dorchester Grey	(* Export only)

MK 1 ZODIAC
Saloons

Dorchester Grey and Bristol Fawn***	Dorchester Grey and Winchester Blue***
Dorchester Grey and Canterbury Green***	Black****
	(*** Two-tone)
	(**** Special order)

The most popular home-market Mk 1 colours appear to have been Canterbury Green, Winchester Blue, Dorchester Grey, Bristol Fawn and Black. Very occasionally an export colour appeared on a home-market car.

Mk 2 CONSUL, ZEPHYR and ZODIAC
All models February 1956 to October 1957

Black	Carlisle Blue
Ivory	Hereford Green
Wells Fawn	Warwick Green
Sarum Blue	Corfe Grey

Colours introduced from October 1957 onwards were identified by code letters on the vehicle trim identification plate:

A	Black		
AA	Richmond Blue	BD	Ming Yellow
AB	Pembroke Coral	BE	Chateau Grey
AC	Newark Grey	BH	Caribbean Turquoise
AD	Ludlow Green	BJ	Linden Green
AE	Durham Beige	BL	Ascot Grey
AF	Arundel Lilac†	BM	Windsor Grey
AG	Guildford Blue	BN	Sapphire Blue
AH	Conway Yellow†	BP	Panama Yellow
AN	Rougemont Red†	BR	Goodwood Green
AP	Cirrus White	BS	Monaco Red
AQ	Morocco Beige	E	Dorchester Grey
AR	Pompadour Blue	M	Norwich Blue
AS	Lichen Green	N	Ambassador Blue
AU	Smoke Grey	W	Brecon Grey
AV	Shark Blue	X	Rialto Red
AX	Vulcan Grey	Y	Dover White
AZ	Imperial Maroon	Z	Kenilworth Blue
BA	Ermine White		
BB	Regency Grey		
BC	Lime Green		

(† Code AF later changed to Arundel Grey; AH to Sunburst Yellow; AN to Monza Red)

APPENDIX C
Production figures

Mk 1 MODELS

Consul	**1951**	**1952**	**1953**	**1954**	**1955**	**1956**	**Total**
Built up	19,787	22,611	32,452	34,998	42,186	5,978	158,012
Knocked down	15,880	13,681	15,773	14,895	13,040	200	73,469
Total	35,667	36,292	48,225	49,893	55,226	6,178	231,481

Mk 1 Consul production ran from January 1, 1951 to February 22, 1956.

Zephyr/Zodiac	**1951**	**1952**	**1953**	**1954**	**1955**	**1956**	**Total**
Built up	2,513	12,656	25,180	27,814	28,075	2,148	98,386
Knocked down	950	11,670	18,950	21,480	22,875	1,000	76,925
Total	3,463	24,326	44,130	49,294	50,950	3,148	175,311

Mk 1 Zephyr/Zodiac production ran from February 12, 1951 to February 22, 1956. Total production was split as follows: Zephyr Six 152,677; Zephyr Zodiac 22,634.

All quoted Mk 1 production figures include convertibles and estate cars, which are believed to account for less than 2% of the total.

Mk 2 MODELS

Consul	**1956**	**1957**	**1958**	**1959**	**1960**	**1961**	**1962**	**Total**
Built up	28,054	42,038	58,853	57,038	54,354	44,265	6,349	290,951
Knocked down	11,990	12,890	8,415	11,368	9,660	4,400	570	59,293
Total	40,044	54,928	67,268	68,406	64,014	48,665	6,919	350,244

Mk 2 Consul production ran from February 1956 to April 1962.

Zephyr/Zodiac	**1956**	**1957**	**1958**	**1959**	**1960**	**1961**	**1962**	**Total**
Built up	23,233	26,986	26,686	36,386	37,195	23,786	3,313	177,585
Knocked down	15,022	22,360	26,470	27,700	18,820	11,835	1,625	123,832
Total	38,255	49,346	53,156	64,086	56,015	35,621	4,938	301,417

Mk 2 Zephyr/Zodiac production ran from January 26, 1956 to April 1962.

Zodiac models accounted for approximately 80,000 of total Zephyr/Zodiac production.

Both Consul and Zephyr/Zodiac figures include convertibles, which accounted for approximately 2% of total production.

Estate car	1956	1957	1958	1959	1960	1961	1962	Total
Consul/Zephyr/Zodiac								
Built up	119	870	898	1,204	1,475	1,039	38	5,643
Pick-up/Utility								
Built up (Nov 56 - Aug 59)	1	13	21	11	—	—	—	46
Knocked down (June 57 - Jan 61)	—	3,030	5,820	4,650	4,030	50	—	17,580
Station wagon								
Australian								
Knocked down (Mar 59 - Jan 61)	—	—	—	3,880	3,530	60	—	7,470

Figures for pick-up/utility and station wagon are for 4-cylinder and 6-cylinder models combined.

APPENDIX D
Major competition awards

1952		
Dutch International Tulip Rally		
Outright winner	Consul Mk 1	K. Wharton/J. Langalaan
1953		
Monte Carlo Rally		
Outright winner	Zephyr Six Mk 1	M. Gatsonides/P. Worledge
1954		
RAC International Rally		
3rd overall	Zephyr Six Mk 1	T. C. Harrison/E. Harrison
1st Saloon car category	Zephyr Six Mk 1	T. C. Harrison/E. Harrison
Dutch International Tulip Rally		
1st 1,301-1,600cc Touring cars	Consul Mk 1	J. G. Reece/P. Reece
2nd 1,301-1,600cc Touring cars	Consul Mk 1	J. W. Fleetwood/G. Read
1st 1,601-2,600cc Touring cars	Zephyr Six Mk 1	R. W. Philips/D. G. Scott
Scottish International Rally		
1st 1,301-1,600cc Touring cars	Consul Mk 1	D. G. Morley/E. Morley
1st 1,601-2,600cc Touring cars	Zephyr Six Mk 1	P. S. Hughes/
1955		
Monte Carlo Rally		
4th overall	Zephyr Six Mk 1	G. N. Burgess/P. Easton
Stuart Trophy (1st British car/driver)	Zephyr Six Mk 1	G. N. Burgess/P. Easton
RAC International Rally		
1st 1,301-2,600cc GT cars	Zephyr Six Mk 1	T. C. Harrison/E. Harrison
2nd 1,301-2,600cc GT cars	Zephyr Six Mk 1	S. H. Allard/T. L. Allard
3rd 1,301-2,600cc GT cars	Zephyr Six Mk 1	J. G. Reece/P. B. Reece
Dutch International Tulip Rally		
1st 1,301-3,500cc Special Series	Zephyr Six Mk 1	T. C. Harrison/R. Habershon
2nd 1,301-3,500cc Special Series	Consul Mk 1	J. G. Reece/P. B. Reece
3rd 1,301-3,500cc Special Series	Zephyr Six Mk 1	S. Henson/Weich
East African Safari		
1st overall	Zephyr Six Mk 1	V. Preston/D. P. Marwaha
Lady McMillan Trophy (Ladies prize)	Zephyr Six Mk 1	Mary Wright
1956		
Alpine Rally		
Coupe des Alpes	Zephyr Mk 2	T. C. Harrison/E. Harrison
Coupe des Alpes	Zephyr Mk 2	D. G. Scott/S. Astbury
1st 2,000-2,600cc category	Zephyr Mk 2	T. C. Harrison/E. Harrison
2nd 2,000-2,600cc category	Zephyr Mk 2	D. G. Scott/S. Astbury
1957		
Dutch International Tulip Rally		
Team prize	Zephyr Mk 2	D. G. Scott/J. D. Irlam
	Zephyr Mk 2	R. J. Adams/E. T. McMillan
	Zephyr Mk 2	T. C. Harrison/J. F. Harrison
Ladies prize	Zephyr Mk 2	Anne Hall/Mary Hopkinson

1958		
Monte Carlo Rally		
1st Over 2,000cc Touring cars	Zodiac Mk 2	R. Nelleman/M. Skarring
RAC International Rally		
1st Over 2,000cc GT cars	Zephyr Mk 2	D. G. Scott/K. Armstrong
East African Safari		
1st overall (see text)	Zephyr Mk 2	A. T. Kopperud/K. M. Kopperud
Team prize (see text)	Zephyr Mk 2	A. T. Kopperud/K. M. Kopperud
	Zephyr Mk 2	F. Brown/F. Collis
	Zephyr Mk 2	V. Preston/K. Springer
Ladies prize	Zephyr Mk 2	Mary Wright
Scottish International Rally		
1st 1,601-2,600cc Touring cars	Zephyr Mk 2	E. Jackson/
Alpine Rally		
Coupe des Alpes	Zephyr Mk 2	E. Harrison/R. Habershon
1st Over 1,600cc Touring cars	Zephyr Mk 2	E. Harrison/R. Habershon
2nd Over 1,600cc Touring cars	Zephyr Mk 2	T. C. Harrison/J. F. Harrison
1959		
East African Safari		
2nd overall	Zephyr Mk 2	D. G. Scott/P. R. Davies
3rd overall	Zephyr Mk 2	E. Harrison/D. L. Markham
Team prize	Zephyr Mk 2	D. G. Scott/P. R. Davies
	Zephyr Mk 2	E. Harrison/D. L. Markham
	Zodiac Mk 2	F. C. Young/L. Baillon
Dutch International Tulip Rally		
3rd overall	Zephyr Mk 2	P. Riley/R. Bensted-Smith
1st 1,601-2,600cc Touring cars	Zephyr Mk 2	P. Riley/R. Bensted-Smith
3rd 1,601-2,600cc Touring cars	Zephyr Mk 2	T. C. Harrison/J. F. Harrison
Team prize	Zephyr Mk 2	P. Riley/R. Bensted-Smith
	Zephyr Mk 2	T. C. Harrison/J. F. Harrison
	Zephyr Mk 2	G. N. Burgess/S. Croft-Pearson
Alpine Rally		
Coupe des Alpes	Zephyr Mk 2	P. Riley/A. Pitts
Coupe des Alpes	Zephyr Mk 2	T. C. Harrison/J. F. Harrison
Coupe des Alpes	Zephyr Mk 2	E. Harrison/W. Fleetwood
1st Over 2,000cc Touring cars	Zephyr Mk 2	P. Riley/A. Pitts
2nd Over 2,000cc Touring cars	Zephyr Mk 2	T. C. Harrison/J. F. Harrison
3rd Over 2,000cc Touring cars	Zephyr Mk 2	E. Harrison/W. Fleetwood
Team prize	Zephyr Mk 2	P. Riley/A. Pitts
	Zephyr Mk 2	T. C. Harrison/J. F. Harrison
	Zephyr Mk 2	E. Harrison/W. Fleetwood
RAC International Rally		
Outright winner	Zephyr Mk 2	G. N. Burgess/S. Croft-Pearson
BRSCC Saloon Car Championship		
Outright winner	Zephyr Mk 2	J. M. Uren
1960		
Monte Carlo Rally		
1st Over 2,000cc GT cars	Zodiac Mk 2	L. Handley/D. Harvey
East African Safari		
3rd overall	Zephyr Mk 2	V. Preston/J. F. Harrison
Team prize	Zephyr Mk 2	V. Preston/J. F. Harrison
	Zephyr Mk 2	T. C. Harrison/P. Davies
	Zephyr Mk 2	D. G. Scott/L. Baillon
1961		
Monte Carlo Rally		
1st Over 2,000cc GT cars	Zodiac Mk 2	L. Handley/D. Harvey
2nd Over 2,000cc GT cars	Zephyr Mk 2	G. N. Burgess/S. Croft-Pearson
3rd Over 2,000cc GT cars	Zephyr Mk 2	E. Jackson/J. Foster
East African Safari		
3rd overall	Zephyr Mk 2	Anne Hall/Lucille Cardwell
Ladies prize	Zephyr Mk 2	Anne Hall/Lucille Cardwell
Team prize	Zephyr Mk 2	Anne Hall/Lucille Cardwell
	Zephyr Mk 2	T. Wisdom/P. Walker
	Zephyr Mk 2	T. C. Harrison/P. Davies

APPENDIX E
Lubrication schedule

The manufacturers recommended the following lubrication procedures as part of the routine servicing of Consuls, Zephyrs and Zodiacs:

Every 1,000 miles
Inject two or three strokes of grease into: track control arm supports; track rods; steering idler arm; steering gear arm rod; gear-change lever; handbrake linkage. Inject SAE 140 oil into universal joints. Top up steering box and gearbox with EP oil. Top up rear axle with Hypoid oil.

Every 5,000 miles
Drain and refill engine sump. Renew oil filter element. Drain and refill gearbox with EP oil. Drain and refill rear axle with Hypoid oil. Pack front hubs with grease. Top up front shock absorber struts and rear shock absorbers with Shock Absorber fluid.

APPENDIX F
Clubs and spares specialists

Owners' clubs
Mk 1 Consul Zephyr Zodiac Owners' Club
General Secretary, Dick Barry
25 Charlwood Gardens
Burgess Hill
Sussex RH15 0RD

Mk 2 Consul Zephyr Zodiac Owners' Club
Secretary, Dave Debenham
26 Burwash Road
Plumstead
London SE18 7QZ

Consul Zephyr Zodiac Club of New South Wales
Secretary, Dave Barry
33 Terry Street
Sydenham
NSW 2044
Australia

The Zephyr and Zodiac Owners' Club of Melbourne
Secretary, Ian Cleary
PO Box 86
Surrey Hills
Victoria
Australia

Consul Zephyr Zodiac Car Club of Queensland
15 Montrebe Street
Holland Park West
Queensland
Australia

Consul Zephyr Zodiac Club of New Zealand
Mark Dunn
37 Jessie Street
Wellington
New Zealand

Consul Zephyr Zodiac Club of Christchurch New Zealand
Grant Lousich
20 Sawyers Arms Road
Papanui
Christchurch 5
New Zealand

The Auckland Consul Zephyr and Zodiac Car Club
Greg Price
c/o The Post Office
Paremoremo
Auckland
New Zealand

Spares Specialists
Ken Tingey, Ford 50 Spares
69 Jolliffe Road
Poole
Dorset BH15 2HA

Newford Parts Centre
Triggs Farm
Fishwick Lane
Higher Wheelton
Nr Chorley
Lancashire